Introduction to Chemicals from Biomass

Wiley Series
in
Renewable Resources

Series Editor

Christian V. Stevens, Department of Organic Chemistry, Ghent University, Belgium

Titles in the Series

Wood Modification: Chemical, Thermal and Other Processes
Callum A. S. Hill

Renewables-Based Technology: Sustainability Assessment
Jo Dewulf & Herman Van Langenhove

Introduction to Chemicals from Biomass
James H. Clark & Fabien E.I. Deswarte

Forthcoming Titles

Biofuels
Wim Soetaert & Erick Vandamme

Starch Biology, Structure & Functionality
Anton Huber & Werner Praznik

Handbook of Natural Colorants
Thomas Bechtold & Rita Mussak

Surfactants from Renewable Resources
Mikael Kjellin & Ingegärd Johansson

Industrial Applications of Natural Fibres: Structure, Properties and Technical Applications
Jorg Müssig

Thermochemical Processing of Biomass
Robert C. Brown

Bio-based Polymers
Martin Peter & Telma Franco

Introduction to Chemicals from Biomass

Editors

JAMES H. CLARK
Green Chemistry Centre, University of York, UK

With

FABIEN E. I. DESWARTE
Green Chemistry Centre, University of York, UK

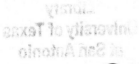

A John Wiley and Sons, Ltd, Publication

Library of Congress Cataloging-in-Publication Data

Introduction to chemicals from biomass / editor, James H. Clark with Fabien E.I. Deswarte.
     p.   cm.
Includes bibliographical references and index.
ISBN 978-0-470-05805-3 (cloth : alk. paper)
1. Biomass chemicals.   2. Organic compounds.   I. Clark, James H.   II. Deswarte, Fabien E. I.
TP248.B55.I68   2008
661'.8—dc22
2008010998

British Library Cataloguing in Publication Data

A catalogue record for this book is available from the British Library

ISBN 978-0-470-05805-3

Typeset in 10/12pt Times by Aptara Inc., New Delhi, India
Printed and bound in Great Britain by TJ International, Padstow, Cornwall

Contents

Series Preface

Renewable resources, their use and modification are involved in a multitude of important processes with a major influence on our everyday lives. Applications can be found in the energy sector, chemistry, pharmacy, the textile industry, paints and coatings, to name but a few.

The area interconnects several scientific disciplines (agriculture, biochemistry, chemistry, technology, environmental sciences, forestry, ...), which makes it very difficult to have an expert view on the complicated interaction. Therefore, the idea to create a series of scientific books, focussing on specific topics concerning renewable resources, has been very opportune and can help to clarify some of the underlying connections in this area.

In a very fast changing world, trends are not only characteristic for fashion and political standpoints, also science is not free from hypes and buzzwords. The use of renewable resources is again more important nowadays, however, it is not part of a hype or a fashion. As the lively discussions among scientists continue about how many years we will still be able to use fossil fuels, opinions ranging from 50 years to 500 years, they do agree that the reserve is limited and that it is essential not only to search for new energy carriers but also for new material sources.

In this respect, renewable resources are a crucial area in the search for alternatives for fossil-based raw materials and energy. In the field of the energy supply, biomass and renewable-based resources will be part of the solution alongside other alternatives such as solar energy, wind energy, hydraulic power, hydrogen technology and nuclear energy.

In the field of material sciences, the impact of renewable resources will probably be even bigger. Integral utilisation of crops and the use of waste streams in certain industries will grow in importance leading to a more sustainable way of producing materials.

Although our society was much more (almost exclusively) based on renewable resources centuries ago, this disappeared in the Western world in the 19th century. Now it is time to focus again on this field of research. However, it should not mean a "retour à la nature", but it should be a multidisciplinary effort on a highly technological level to perform research towards new opportunities, to develop new crops and products from renewable resources. This will be essential to guarantee a level of comfort for a growing number of people living on our planet. It is "the" challenge for the coming generations of scientists to develop more sustainable ways to create prosperity and to fight poverty and hunger in the world. A global approach is certainly favoured.

This challenge can only be dealt with if scientists are attracted to this area and are recognized for their efforts in this interdisciplinary field. It is therefore also essential that consumers recognize the fate of renewable resources in a number of products.

Furthermore, scientists do need to communicate and discuss the relevance of their work. The use and modification of renewable resources may not follow the path of the genetic engineering concept in view of consumer acceptance in Europe. Related to this aspect, the series will certainly help to increase the visibility of the importance of renewable resources.

Being convinced of the value of the renewables approach for the industrial world, as well as for developing countries, I was myself delighted to collaborate on this series of books focussing on different aspects of renewable resources. I hope that readers become aware of the complexity, the interaction and interconnections, and the challenges of this field and that they will help to communicate on the importance of renewable resources.

I certainly want to thank the people of Wiley from the Chichester office, especially David Hughes, Jenny Cossham and Lyn Roberts, in seeing the need for such a series of books on renewable resources, for initiating and supporting it and for helping to carry the project to the end.

Last, but not least I want to thank my family, especially my wife Hilde and children Paulien and Pieter-Jan for their patience and for giving me the time to work on the series when other activities seemed to be more inviting.

Christian V. Stevens
Faculty of Bioscience Engineering
Ghent University, Belgium
Series Editor '*Renewable Resources*'
June 2005

Preface

The growth of interest in the 21st century in the biorefinery concept has been dramatic, if somewhat chaotic and disjointed, with interest and activity in the various relevant disciplines ranging from significant to very limited, and with little connection being made between those disciplines. Biologists and bioengineers have been quick to see the opportunity for white biotechnology, the application of biotechnology to industrial production, in particular through the use of whole cells and enzymes to synthesise products. Such applications are not new – biotechnology has been contributing to industrial processes for some time. For decades, bacterial enzymes have been used widely in food manufacturing and as active ingredients in washing powders. Transgenic *Escherichia coli* are used to produce human insulin in large-scale fermentation tanks, and the first rationally designed enzyme, used in detergents to break down fat, was introduced as early as 1988.

The involvement of other disciplines in the development of biorefineries has been less apparent so far, at least as judged by the literature and attendance at international symposia and other events. Yet, it can be argued that experts in subjects including environmental studies and ecology, social policy and economics, and chemistry and process engineering are all needed to take forward this most essential of societal technological developments. Furthermore, these disciplines need to learn how to work together: interdisciplinarity and multidisciplinarity are the order of the day, including collaboration between technical and non-technical experts and organisations. While white biotechnology is proven, its scope is limited and cannot be expected to answer a very high proportion of the challenges the biorefineries of the future will face. The belief among some early 21st century biologists that by thinking in terms of small molecules they can manage without chemists, is as naïve as the belief among some early 20th century chemists that by thinking in terms of atoms, they can do without physicists.

The reality is that chemists need biologists to learn how to harness the power of biological systems to engineer molecules, while biologists need chemists to understand what to use bioengineering for and how to build up the molecules that it produces to valuable products. Process and chemical engineers are needed to make it practical for industry to actually make the products in the quantities needed and at the rate of production required. But perhaps unlike any previous step change in technology, for the sustainable exploitation of biomass that is so intrinsically linked to the wellbeing of the environment and the survival of life, the scientists and technologists need additional expertise to ensure that we really do achieve economic, environmental and societal benefit, and here we must engage experts from non-technical disciplines.

We have a wonderful opportunity to learn from the mistakes, as well as the successes, of close to a century of petroleum exploitation, two centuries of industrialization and many more centuries of living off the land. These must all come together and work together in a sustainable manner if our standards of living in the developed world are to be maintained and if those of the developing world are to grow. How are we to balance the need for greater consumption that the expectations of the developing world seems to require with diminishing traditional resources and the alarming level of pollution to land, sea and air that our consumption of those resources inexorably creates? In Chapter 1 of this book we look more closely at this apparent dichotomy and argue the case for transmaterialisation – the fundamental change in the resources that we consume (e.g. from fossil to biomass), alongside a massive reduction in the cradle-to-cradle cycle time of these resources, and all done with a much increased level of awareness of ecosystem services. Put these together and we have the basis for future sustainable biorefineries – small and large, using various feedstocks and in various locations, but all producing products that cause neither our generation nor future generations unreasonable stresses on the Earth.

As biorefinery research has been dominated thus far by biotechnology, biorefinery production has been heavily focused on energy products. In fact biorefineries must produce many different products, analogous to petroleum refineries – energy, chemicals and materials are all needed by today's society and in increasing amounts. In Chapter 2 David Turley explores the chemical value of biomass. This includes plant oils, a topic of enormous current interest, due to their value in chemicals as well as fuel, but also as their over-exploitation can lead to major local environmental damage that some believe may even make their overall impact unfavourable. The largest renewable sources of carbon are the carbohydrates and the, currently limited, non-food uses for starch, cellulose and hemicellulose are considered here. Other important biomass components including lignin, waxes and proteins are also reviewed.

The conversion of this enormous natural chemical potential to the actual products we want requires chemical technologies and if we are to keep the overall environmental footprint low and build on the head start afforded by renewable feedstocks, these need to be green chemical technologies. Fran Kerton reviews these in Chapter 3. Apart from the tools of biotechnology, some of the key technologies are likely to be alternative solvents – for extraction of valuable plant chemicals and for chemical processing; alternative activation methods including microwaves (so routinely used by millions of people to convert biomass into food, but undoubtedly with much wider potential for exploitation), ultrasound, electrochemistry (a much underutilised technology, yet with enormous proven value for molecular synthesis and other material and chemical processing) and photochemistry; and catalysis including environmentally friendly heterogeneous catalytic technologies, as well as some highly selective homogeneous catalysis and of course, biocatalysis, including the modifications that chemists have applied to enhance their green credentials such as immobilization (to aid recovery and reuse) and supercritical fluid reaction media.

Apostolis Koutinas and colleagues bring the chemical potential of biomass and the technologies for its conversion together in Chapter 4. The future bioplatform molecules from which many of the future sustainable chemicals will be made are considered, as are the technologies to make them, including fermentations, enzymatic transformations, and various chemical transformations including simple acid catalysis. Lignocellulosics, proteins and vegetable oils, are considered as feedstocks.

Materials from biomass are the subjects of Chapter 5. Carlos Vaca-Garcia looks at biomaterials and biopolymers from biomass. The scope of this chapter is enormous and starts by looking at cellulose itself and the long-used building and clothing materials of society, notably wood and cotton, as well as others, including sisal, ramie, jute and hemp. The particular importance of cellulose, nature's most abundant material, is shown through a consideration of modifications, both physical and chemical, and the uses for major derivatives, notably cellulose esters. Starch derivatives, chitin and its derivative chitosan are among other materials considered. The latter parts of this chapter include a study of biodegradable plastics including PLA, probably the best known 'green' plastic at the time of writing.

The chemical value of a refinery can match that of the fuels: the value of the smaller volume, but higher price chemical components have kept petrorefineries economically viable for close to a century. However, the chemical–energy relationship is symbiotic: the chemicals also need the energy. Most of the new biorefineries of Europe and the Americas have one or more energy products at their heart – most commonly biodiesel and bioethanol, but soon extending to others. The bulk of the biomass carbon will end up in a fuel product and to ensure the overall sustainability and low environmental footprint of the total biorefinery we need to work hard to maximize the energy efficiency, and this is discussed from a largely chemical viewpoint by Mehrdad Arshadi and Anita Sellstedt in Chapter 6. Biomass can be used as fuel in many different ways including solid pellets, liquid fuels, hydrogen, biogas and other fuels for the future, and these are all considered here. The raw materials, the main methods of conversion, the efficiency of conversion, the engineering involved and the by-products are all reviewed.

This book is intended to show that while biomass has biological origins and that biological methods will be an important part of the biorefinery toolkit, biomass is essentially a rich mixture of chemicals and materials that can be extracted for use or that can be converted into other useful products including high value chemicals. When we manipulate biomass we must build on the ideal start for any product lifecycle of a renewable resource, and use green (bio)chemical technologies for transformations to what could be genuinely green and sustainable products. We need to bring the chemistry to the biorefinery as well as take the chemicals from the biorefinery – and we must make sure it's all green chemistry!

I would like to express my thanks to my co-editor Fabien Deswarte who took ownership of this project from day one and who tirelessly and patiently worked with each author to nurse them through to completion of their chapters. My thanks also go to all of those authors who accepted their tasks with enthusiasm and who have made such valuable contributions to this book.

James Clark, York, UK
January 2008

List of Contributors

Mehrdad Arshadi   Unit of Biomass Technology and Chemistry / BTK, Swedish University of Agricultural Sciences, Umea, SWEDEN

James H. Clark   Clean Technology Centre, Chemistry Department, University of York, Heslington, York, UK

Fabien E.I. Deswarte   Clean Technology Centre, Chemistry Department, University of York, Heslington, York, UK

C. Du   School of Chemical Engineering and Analytical Science, University of Manchester, Sackville Street, Manchester, UK

Francesca M. Kerton   Department of Chemistry, Memorial University of Newfoundland, St. John's, NL, CANADA

Apostolis A. Koutinas   School of Chemical Engineering and Analytical Science, University of Manchester, Sackville Street, Manchester, UK

Anita Sellstedt   Unit of Biomass Technology and Chemistry / BTK, Swedish University of Agricultural Sciences, Umea, SWEDEN

David B. Turley   Agriculture and Rural Strategy, Central Science Lab, Sand Hutton, York, UK

Carlos Vaca Garcia   ENSIACET Laboratoire de Chimie Agro-Industrielle, 118 route de Narbonne, 31077 Toulouse, FRANCE

R. H. Wang   School of Chemical Engineering and Analytical Science, University of Manchester, Sackville Street, Manchester, UK

Colin Webb   School of Chemical Engineering and Analytical Science, University of Manchester, Sackville Street, Manchester, UK

1

The Biorefinery Concept–An Integrated Approach

James H. Clark and Fabien E. I. Deswarte
Green Chemistry Centre of Excellence, University of York, UK

1.1 The Challenge of Sustainable Development

Reconciling the needs of a growing world population with the resulting impact on our environment is ultimately the most complex and important challenge for society. Sustainable development requires an assessment of the degree to which the natural resources of the planet are both in sufficient quantity and in an accessible state to meet these needs, and to be able to deal with the wastes that we inevitably produce in manipulating these resources (including process and end-of-life waste). We can express this in the form of an equation based on the Earth's capacity EC, the total population exploiting it P, the consumptive (essentially equating to economic) activity of the average person C, and an appropriate conversion factor between activity and environmental burden B.

$$EC = P \times C \times B$$

In a period, such as the present time, of growth in P and C, the latter through economic growth in the developing world, notably India and China, (and an assumption that EC is not limitless and that we may not be far short of reaching its limit) we can only move towards sustainability through a reduction in B.

Introduction to Chemicals from Biomass Edited by James Clark and Fabien Deswarte
© 2008 John Wiley & Sons, Ltd

How can we reduce B? There are only two appropriate routes:

- Dematerialisation (use less resource per person)
- Transmaterialisation (replacement of current raw materials including energy)

Dematerialisation has, to some extent, been a natural part of our technological progress, with less and less resources (e.g. measured as amount of carbon) being used to generate a unit of activity (e.g. measured on the basis of gross domestic product). We have been progressively developing more efficient technologies and legislation and other pressures have forced the processing industries to reduce waste and to make use of that waste through recycling and reuse. However, there are conflicting societal trends that reduce these positive effects on B. Our increasing wealth has brought with it an increase in levels of consumption with individuals using their increasing wealth to buy more goods and to buy more often. The average number of cars, area of housing, quantity of clothes, food purchased and consumer goods (e.g. electronics) per person in the wealthier countries inexorably increases, while the lure of advertising encourages people to change their personal possessions at a rate way above that commensurate with the items' wear and tear.

Transmaterialisation is a more fundamental approach to the problem, which, with the goal of sustainable development, would ultimately switch consumption to only those resources that are renewable on a short timescale. Clearly petroleum, which takes millions of years to form, is not an example of such a sustainable resource. For the method to be truly effective, the wastes associated with the conversion and consumption of such resources must also be environmentally compatible on a short timescale. The use of polyolefin plastic bags for example, which have lifetimes in the environment of hundreds of years, is not consistent with this (no matter how they compare with alternative packaging materials at other stages in their lifecycle), nor is the use of some hazardous process auxiliaries which are likely to cause rapid environmental damage on release into the environment.

While manufacturing processes have largely become more efficient, both in terms of use of resources and in terms of reduced waste, industry needs to regularly and thoroughly monitor its practises through full inventories of all inputs and outputs. Gate-to-gate environmental footprints help to identify hotspots where new technology can make a significant difference, and help to determine the value of any changes made. In chemical processes, green chemistry metrics such as mass intensity and atom efficiency need to be used alongside yield, and companies need to assist their researchers and process chemists by developing in-house guides (e.g. over choice of solvent), assessment methods, and recommended alternative reagents and technologies. In their present form, these mechanisms are, however, largely limited to further steps towards dematerialisation. Progress towards transmaterialisation requires additional features to be taken into consideration and in some cases a very different way of thinking of the problems. We must add the sustainability of all manufacturing components, inputs and outputs. Are the feedstocks for a particular manufacturing process from sustainable sources? Are the process auxiliaries sustainable? Are the process outputs – product(s) and waste – environmentally compatible e.g. through rapid biodegradation (ideally with the waste having a valuable use, even if it is a completely different application, so that the inevitable release into the environment, as is the fate for all materials, is delayed).

For organic chemicals, transmaterialisation must mean a shift from fossil (mainly petroleum) feedstocks (which have a cycle time of $>10^7$ years) to plant-based feedstocks (with cycle times of $<10^3$ years). This immediately raises several fundamentally important questions: Can we produce and use enough plants to satisfy the carbon needs of chemical and related manufacturing, while not compromising other (essentially food and feed) needs? Do we have the technologies necessary to carry out the conversions (biomass to chemicals) and in a way that does not completely compromise the environmental and transmaterialisation characteristics of the new process?

1.2 Renewable Resources — Nature and Availability

We need to find new ways of generating the chemicals, energy and materials, as well as food that a growing world population (increasing 'P') and growing individual expectations (increasing 'C') needs, doing so while limiting environmental damage. At the beginning of transmaterialisation is the feedstock or primary resource, and this needs to be made renewable (see Table 1.1). An ideal renewable resource is one that can be replenished over a relatively short timescale or is essentially limitless in supply. Resources such as coal, natural gas and crude oil come from carbon dioxide 'fixed' by nature through photosynthesis many millions of years ago. They are of limited supply, cannot be replaced and thus are non-renewable. In contrast, resources such as solar radiation, winds, tides and biomass can be considered as renewable resources, which are (if appropriately managed) in no danger of being over-exploited. However, it is important to note that, while the first three resources can be used as a renewable source of energy, biomass can be used to produce not only energy, but also chemicals and materials – the focus of this book.

By definition, biomass corresponds to any organic matter available on a recurring basis (see Figure 1.1). The two most obvious types of biomass are wood and

Table 1.1 *Different types of renewable and non-renewable resources*

Non-Renewable Resources	Renewable Resources
Coal	Sun
Natural gas	Tides and Hydro
Crude oil	Biomass
	Wind

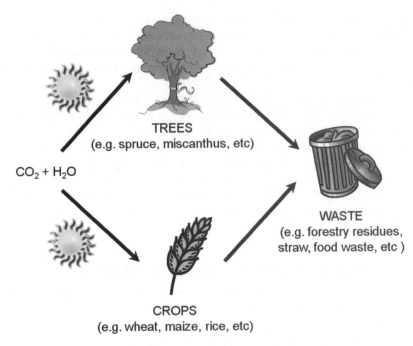

Figure 1.1   *Different types of biomass*

crops (e.g. wheat, maize and rice). Another very important type of biomass we tend to forget is waste (e.g. food waste, manure, etc). These resources are generally considered to be renewable as they can be continually re-grown/regenerated. They take up carbon dioxide from the air while they are growing (through photosynthesis) and then return it to the air at the end of life, thereby creating a 'closed loop' (Deswarte, 2008).

Food crops can indeed be used to produce energy (e.g. biodiesel from vegetable oil), materials (e.g. polylactic acid from corn) and chemicals (e.g. polyols from wheat). However, it is now becoming widely recognised by governments and scientists that waste and lignocellulosic materials (e.g. wood, straw, energy crops) offer a much better opportunity, since they avoid competition with the food sector and, often, do not require as much land and fertilisers to grow. In fact, only 3% of the 170 million tonnes of biomass produced yearly by photosynthesis is currently being cultivated, harvested and used (food and non-food applications) (Sanders *et al.*, 2005). Indeed, according to a recent report published by the USDOE and the USDA (2005), the US alone could sustainably supply more than one billion dry tons of biomass annually by 2030. As seen in Table 1.2, the biomass potential in Europe is also enormous.

About 10% of all the oil we extract in the world is used to make organic chemicals and related materials. A remarkable additional 10% is used for energy to drive the

Table 1.2   Biomass potential in the EU (European Commision, 2006)

	Biomass Potential (MToe)		
	2010	2020	2030
Organic Wastes	100	100	102
Energy Crops	43–46	76–94	102–142
Forest Products	43	39–45	39–72
Total	**186**	**215–239**	**243–316**

chemical reactions. In the EU, this corresponded to 166 million tonnes in 2000. While increases in efficiency of chemical manufacturing in the EU have been considerable, an OECD estimate has shown that the chemical industry worldwide produces about 4% of global carbon dioxide emissions (10^{12} tonnes). A shift away from fossil resources should thus benefit both resource depletion, pollution and global warming.

1.3   Impact on Ecosystem Services

Ecosystem services are the goods and services provided by coupled and ecological social systems. They are at the heart of our quality of life by providing the materials on which we base our lifestyles, and we all inevitably depend on the sustainable use of ecosystem services. The millennium ecosystems assessment brought this to our attention (*Ecosystems and Human Well-Being*, 2007) by stating that the ability of many ecosystems to deliver valuable services has been compromised by resource over-exploitation and by environmental degradation.

The figures provided by the European Biofuels Research Advisory Council (see Table 1.2) suggest an increasing potential for the conversion of biomass to biofuels in Europe over the next 20+ years, but can the European environment cope with ever-increasing biomass exploitation? We must give greater consideration to the associated stresses on large areas of land and associated systems, including water, food production and recreation (even the use of low value/waste materials such as straw and grasses will have effects). In general, when considering such enormous changes in ecosystem services exploitation we need to:

- Study the associated changes in the quality and availability of local ecosystem services
- Consider how activities in one region can affect ecosystem services elsewhere
- Study the linkage between livelihoods, human well-being and ecosystem services
- Consider how to manage the ecosystem services under pressure.

1.4 The Biorefinery Concept

1.4.1 Definition

One way to mitigate the negative effects of local ecosystem services is to convert biomass into a variety of chemicals (Chapters 2 and 4), biomaterials (Chapter 5) and energy (Chapter 6), maximising the value of the biomass and minimising waste. This integrated approach corresponds to the biorefinery concept and is gaining increased attention in many parts of the world (Kamm and Kamm, 2004; Halasz *et al.*, 2005) As illustrated in Figure 1.2, the biorefinery of the future will be analogous to today's petrorefineries (Realff and Abbas, 2004; National Renewable Energy Laboratory, www.nrel.gov/biomass/biorefinery.html).

Similarly to oil-based refineries, where many energy and chemical products are produced from crude oil, biorefineries will produce many different industrial products from biomass. These will include low-value, high-volume products, such as transportation fuels (e.g. biodiesel, bioethanol), commodity chemicals, as well as materials, and high-value, low-volume products or speciality chemicals, such as cosmetics or nutraceuticals. Energy is the driver for developments in this area, but as biorefineries become more and more sophisticated with time, other products will be developed. In some types of biorefinery, food and feed production may well also be incorporated.

1.4.2 Different Types of Biorefinery

Three different types of biorefinery have been described in the literature (van Dyne *et al.*, 1999; Kamm & Kamm, 2004; Fernando *et al.*, 2006):

- Phase I biorefinery (single feedstock, single process and single major product)
- Phase II biorefinery (single feedstock, multiple processes and multiple major products)

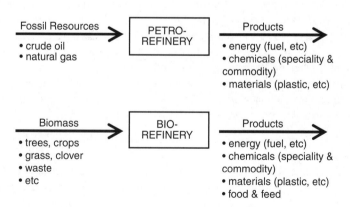

Figure 1.2   *Comparison of petrorefinery vs. biorefinery*

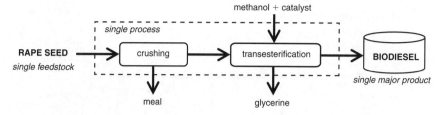

Figure 1.3   *The biodiesel process – an example of a phase I biorefinery*

- Phase III biorefinery (multiple feedstocks, multiple processes and multiple major products).

Phase I Biorefinery

Phase I biorefineries use one only feedstock, have fixed processing capabilities (single process) and have a single major product. They are already in operation and are proven to be economically viable. In Europe, there are now many 'phase I biorefineries' producing biodiesel. They use vegetable oil (mainly rapeseed oil in the EU) as a feedstock and produce fixed amounts of biodiesel and glycerine through a single process called transesterification (see Figure 1.3). They thus have almost no flexibility to recover investment and operating costs. Other examples of phase I biorefinery include today's pulp and paper mills, and corn grain-to-ethanol plants.

Phase II Biorefinery

Similarly to phase I biorefineries, phase II biorefineries can only process one feedstock. However, they are capable of producing various end products (energy, chemicals and materials) and thus respond to market demand, prices, contract obligation and the plant's operating limits. One example of a phase II biorefinery is the Novamont plant in Italy, which uses corn starch to produce a range of chemical products including biodegradable polyesters (Origi-Bi) and starch-derived thermoplastics (Mater-Bi) (www.materbi.com). Another example of this type of biorefinery is the Roquette site at Lestrem in France that produces a multitude of carbohydrate derivatives, including native and modified starches, sweeteners, polyol and bioethanol from cereal grains (www.roquette.fr/index_eng.asp, see Figure 1.4).

Roquette produces more than 600 carbohydrate derivatives worldwide and is now leading a major programme (the BioHub™ programme, supported by the French Agency for Industrial Innovation) aiming to develop cereal-based biorefineries and, in particular, a portfolio of cereals-based platform chemicals (e.g. isosorbide) for biopolymers, as well as speciality and commodity chemicals production (www.biohub.fr).

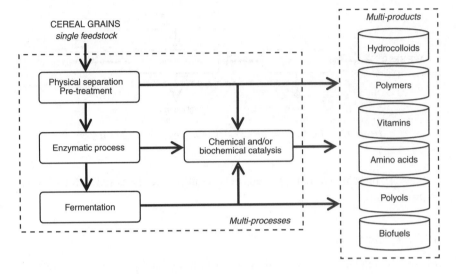

Figure 1.4   *Roquette – an example of a phase II biorefinery (Based on Rupp-Dahlem, 2006)*

Ultimately, all phase I biorefineries could be converted into phase II biorefineries, if we can identify ways to upgrade the various side streams. A phase I biodiesel processing plant, for example, could turn into a phase II biorefinery, if we can develop technologies that can convert biodiesel glycerine (crude glycerol) into valuable energy and chemical products (see Chapter 4 for potential chemical products from glycerol). In fact, it is recognised that energy or biofuel generation will probably (at first) form the 'back bone' of numerous phase II biorefineries, due to large market demand. Interestingly, crude oil refining also started with the production of energy, and has ended up employing sophisticated process chemistry and engineering to develop complex materials and chemicals that 'squeeze every ounce of value' from a barrel of oil (Realff and Abbas, 2004).

Phase III Biorefinery

Phase III biorefineries correspond to the most developed/advanced type of biorefinery. They are not only able to produce a variety of energy and chemical products (phase II biorefineries), but can also use various types of feedstocks and processing technologies to produce the multiplicity of industrial products our society requires. The diversity of the products gives a high degree of flexibility to changing market demands (a current by-product might become a key product in the future) and provides phase III biorefineries with various options to achieve profitability and maximise returns. In addition, their 'multiple feedstock' aspect helps them to secure feedstock availability and offers these highly integrated biorefineries the possibility to select the most profitable combination of raw materials (de Jong *et al.*, 2006). Although no commercial phase III biorefineries exist at present, extensive work is being carried out in the EU (e.g. Biorefinery Euroview, BIOPOL, SUSTOIL), the US (the present leading player in this field) and elsewhere on the design and feasibility of such facilities. Full-scale phase III (zero-waste) biorefineries are probably more than a decade away – according to a recent report from the Biofuels Research Advisory Council, large integrated biorefineries are not expected to become established in Europe until around 2020 (European Commision, 2006).

Currently, there are four phase III biorefinery systems being pursued in research and development, which will be discussed in more detail in this chapter:

- Lignocellulosic feedstock biorefinery
- Whole crop biorefinery
- Green biorefinery
- Two-platform concept biorefinery.

Lignocellulose feedstock biorefinery   A lignocellulose feedstock biorefinery will typically use 'nature-dry' lignocellulosic biomass such as wood, straw, corn stover, etc. The lignocellulosic raw material (consisting primarily of polysaccharides and lignin) will enter the biorefinery and, through an array of processes, will be fractionated and converted into a variety of energy and chemical products (see Figure 1.5).

The Icelandic Biomass Company is currently running a demonstration plant processing 20 000 tonnes per year of lignocellulosic biomass, including hay, lupine straw and barley straw (Kamm and Kamm, 2005). The plant can produce up to 7 million litres of ethanol per year, and a variety of chemical products from lignin and the various side streams. The University of York, in collaboration with a number of industrial partners, also demonstrated that supercritical CO_2 could be used – as an initial stage in a biorefinery – to extract high value wax products (e.g. nutraceuticals, insect repellents) from straw prior to converting the lignocellulosic

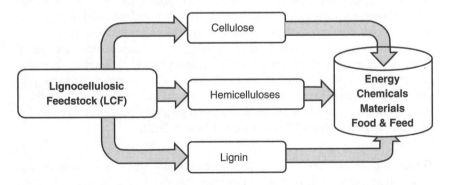

Figure 1.5   *Simplified schematic diagram of a lignocellulosic feedstock biorefinery*

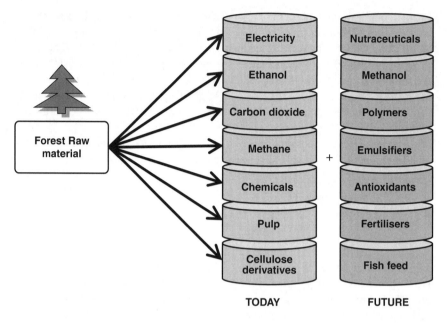

Figure 1.6   *Processum – an example of lignocellulosic feedstock biorefinery*

fraction into paper, strawboard, high quality mulch or energy (Deswarte *et al.*, 2007).

Another example of an imminent lignocellulosic feestock biorefinery is Processum in Sweden, which corresponds to an integrated cluster of industries converting wood into energy, and different chemicals and materials (see Figure 1.6). This is probably one of the best examples of industrial symbiosis – one industry uses the waste of another as a raw material (Gravitis, 2006). Amongst the member companies are Nobel Surface Chemistry (production of thickeners for water-based paints and the construction industry), Domsjo Fabriker (production of global scale dissolving pulp and paper pulp), Ovik Energy (energy production and distribution) and Sekab (production of ethanol, ethanol derivatives and ethanol as fuel).

In reality, while the sole products of existing pulp and paper manufacturing facilities today are pulp and paper (phase I biorefinery), these facilities are geared to collect and process substantial amounts of lignocellulosic biomass. They thus provide an ideal foundation to develop advanced lignocellulose feedstock biorefineries. Additional processes could be built around pulp mills, either as an extension or as an 'across-the-fence'-type company (Agenda 2020).

Whole crop biorefinery   A whole crop biorefinery will employ cereals (e.g. wheat, maize, rape, etc) and convert the entire plant (straw and grain) into energy, chemicals and materials (see Figure 1.7).

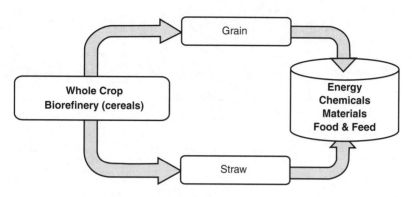

Figure 1.7   *Simplified schematic diagram of a whole crop biorefinery*

The first step will be to separate the seed from the straw (collection will obviously occur simultaneously, to minimise energy use and labour cost). The seeds may then be processed to produce starch and a wide variety of products, including ethanol and bioplastics (e.g. polylactic acid). The straw can be processed to products via various conversion processes, as described above for a lignocellulosic feedstock biorefinery.

POET (formerly known as Broin Companies), the current largest producer of ethanol in the world, are currently building a commercial whole crop biorefinery in Iowa, with a completion date expected in 2011 (www.poetenergy.com). Through the 'LIBERTY project' (jointly funded by POET and the US Department of Energy), a grain-to-ethanol plant (Voyager Ethanol), will be converted from a 50 million gallon per year conventional corn dry mill facility into a 125 million gallon per year commercial-scale biorefinery designed to utilise advanced corn fractionation and lignocellulosic conversion technology to produce ethanol from corn fibre and corn stover (see Figure 1.8). The facility will also produce a number of valuable product, including corn germ and a protein-rich dried distillers grain (Dakota Gold® HP or DGHP), which can be used as an animal feed.

Green biorefinery   Green biorefinery is another form of phase III biorefinery that has been extensively studied in the EU (especially Germany, Austria and Denmark) over the last decade. It takes 'natural wet' green biomass (such as green grass, lucerne, clover, immature cereals, algae, etc.) and converts it into useful products including energy, chemicals, materials and feed, through the use of a combination of different technologies, including fermentation (see Figure 1.9). Typically, green biomass is separated into a fibre-rich press cake and a nutrient-rich green juice (Andersen and Kiel, 2000). The green juice contains a number of useful chemicals such as amino acids, organic acids and dyes. The press cake can be used for fodder or to produce energy, insulation materials, construction panels, biocomposites, etc.

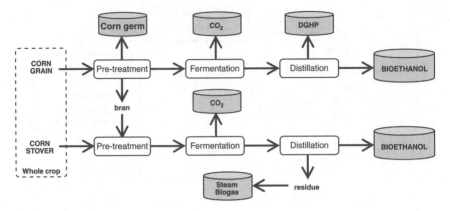

Figure 1.8   *The LIBERTY project – an example of a whole crop biorefinery (Based on Broins Companies, 2007)*

A green biorefinery demonstration plant has recently been set up in Brandenburg (Germany) and produces high-value proteins, lactic acid and fodder from 30 000 tonnes per year of alfalfa and wild mix grass. Another example of this form of phase III biorefinery is the Austrian Green Biorefinery, which is based on the processing of green biomass from silage (see Figure 1.10). A demonstration facility, which is currently being built in Utzenaich, will produce a range of chemicals (e.g. lactic acid, amino acids) and fibre-derived products (e.g. animal feed, boards, insulation materials, etc.) as well as electricity and heat (it will be attached to an existing biogas plant).

Two-platform concept biorefinery   Another form of biorefinery, which has been recently defined by the National Renewable Energy Laboratory (NREL) (www.nrel.gov/biomass/biorefinery.html) is the two-platform concept biorefinery.

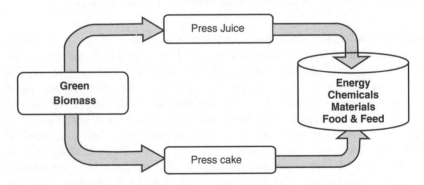

Figure 1.9   *Simplified schematic diagram of a green biorefinery*

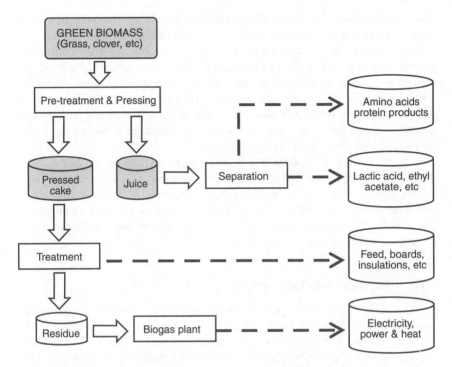

Figure 1.10   *A simplistic schematic of the 'Austrian Green Biorefinery' (Based on Schnitzer, 2006)*

As depicted in Figure 1.11, the feedstock is separated into a 'sugar platform' (bio-chemical) and a 'syngas platform' (thermochemical). Both platforms can offer energy, chemicals, materials, and potentially food and feed, and thus make use of the entire feedstock(s). The 'sugar platform' is based on biochemical conversion processes and focuses on the fermentation of sugars extracted from biomass feed-stocks. The 'syngas platform' thermolytically transforms biomass into gaseous or

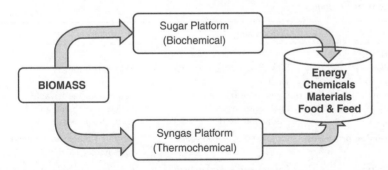

Figure 1.11   *Simplified schematic diagram of a two-platform concept biorefinery*

liquid intermediate chemicals that can be upgraded to transportation fuels, as well as commodity and speciality chemicals (Wright and Brown, 2007a).

No biorefinery of this type currently exists in Europe, but sugar conversion technologies (e.g. wheat grain-to-ethanol fermentation) and the gasification approach (e.g. Choren's Carbo-V® process) are independently used (NNFCC, www.nnfcc.co.uk). Opinions vary widely on the best strategy to combine these two platforms. However, it is most likely that multiple biorefinery designs will emerge commercially – as new technologies are developed – depending on the location of the plant and the feedstock(s) used.

It is interesting to note that the sugar platform and many other (non-thermochemical) processes likely to be incorporated into a biorefinery, will almost certainly generate some waste products that will be difficult to convert into value-added materials and chemicals. Such wastes and residues represent an important source of energy within the biorefinery and are an ideal candidate for thermochemical conversion (Ragauskas *et al.*, 2006).

1.4.3   Challenges and Opportunities

Biorefinery products (energy, chemicals and materials) will most likely have to compete with existing and future petroleum-derived products. As seen in Table 1.3 (comparison of biorefinery and petrorefinery characteristics in terms of feedstock, process and products), the two types of refinery display major differences, which translate into a number of challenges and opportunities to the deployment of biorefineries.

Feedstock

In contrast to fossil resources, which are found in rich deposits ('mine mouths' or 'well heads'), biomass is widely distributed geographically (multiplicity of 'farm

Table 1.3   *Comparison of petrorefinery and biorefinery in terms of feedstock, conversion processes and products*

	Petrorefinery	Biorefinery
Feedstock:		
Location	Rich deposits in some areas	Widely distributed
Density	High	Low
Availability	Continuous but finite	Seasonal but renewable
Chemical composition	Hydrocarbons, not functionalised	Highly oxygenated and functionalised
Conversion processes	Optimised over 100 years	Require further research and technological development
Products	On the market and to high specification	Quality needs to be standardised

gates') (Gravitis, 2007; Wright and Brown, 2007b). In addition, biomass typically exhibits a low bulk density and a relatively high water content (up to 90% for grass), which makes its transport much more expensive than the transfer of natural gas or petroleum.

Reducing the cost of collection, transportation and storage of biomass through densification is thus critical to developing a sustainable infrastructure capable of working with significant quantities of raw material (Hess *et al.*, 2003). In addition, the economics of many conversion processes, which are batch operations, would be dramatically improved through an increase in density, as the inherent low density of biomass limits the amount of material that can be processed at any one time. The most common strategy used to increase biomass density is grinding. By chopping bailed straw, for example, a 10-fold densification can typically be achieved. An alternative strategy, which can provide a material of even higher density, is pelletisation (see Chapter 6 for more information). Through conversion of ground straw into pellets, the density of the material can be further increased by a factor of three (Deswarte *et al.*, 2007). This pre-treatment also offers the added benefit of providing a much more uniform material (in size, shape, moisture, density and energy content), which can be much more easily handled. Pre-processing might be done 'on the farm', but can also be done during harvesting. An example of technology recently developed to address the engineering challenge presented by low bulk density biomass such as wheat straw, is a multicomponent harvester, which can simultaneously and selectively harvest wheat grain and the desired parts of wheat straw in a single pass (Hess *et al.*, 2003).

Another issue associated with the use of (fresh) biomass is its perishable character or susceptibility to degradation. Taking straw (again) as an example, fermentation will begin if the moisture content of baled straw is kept above 25% for a prolonged period of time, resulting in a dramatic reduction in the quality of the raw material. In some cases, spontaneous combustion in the stacks can even take place (Kadam *et al.*, 2000). This issue is particularly important given that, in contrast to fossil resources (which are of permanent availability – continuously pumped and mined), the availability of biomass is seasonal (Thorsell *et al.*, 2004). Thus, in order to ensure a continuous all year round operation of the biorefinery, biomass may have to be stabilised (e.g. dried) prior to (long-term) storage. The Austrian Green Biorefinery, for example, tackles this problem by processing not only direct-cut grasses, but also silage, which can be prepared in the growing season and stored in a silo (Koschuh *et al.*, 2005; Thang and Novalin, 2007).

In summary, it is essential that we develop a cost-effective infrastructure for production, collection, storage and pre-treatment of biomass. As highlighted by Nilsson and Kadam, the economic success of a large biorefinery will greatly depend upon the fundamental logistics of a consistent and orderly flow of feedstocks. (Nilsson, 1999; Kadam *et al.*, 2000). Localised small-scale (and perhaps mobile) pre-treatment units will be necessary to minimise transportation costs and supply the biorefinery with a 'stabilised' feedstock (e.g. in the form of a dry solid or a liquid (pyrolysis oil)), which can be stored and thus allow the biorefinery to run continuously all year long (Sanders *et al.*, 2005). Such an approach will present the added benefits of reducing the environmental impact of transportation (Koschuh *et al.*, 2005) and allowing farmers to gain a greater share of the total added value of the supply chain.

Conversion Processes

The major impediment to biomass use is the development of economically viable methods (physical, chemical, thermochemical and biochemical) to separate, refine and transform it into energy, chemicals and materials (European Commission, 2005). Indeed, biorefining technologies (some of which are already at a stage of commercialisation, while others require further research and technological development) have to compete with processes that have been continuously improved by petrorefineries over the last 100 years (and have a very high degree of technical and cost optimisation). In particular, biorefineries will have to develop clever process engineering to deal with separation – by far the most wasteful and expensive stage of biomass conversion, and currently accounting for 60 to 80% of the process cost of most mature chemical processes. The production of chemicals (e.g. succinic acid) and fuels (e.g. bioethanol) through fermentation processes, for example, generates very dilute and complex aqueous solutions, which will have to be dealt with using clean and low-energy techniques. In fact, given that so many of our carefully isolated, functionalised and purified chemical products end up in formulations, it would also seem wise to seek methods that can convert the multicomponent systems we obtain from biomass into multicomponent formulations with the correct set of properties we require in applications such as cleaning, coating and dyeing (Clark, 2007). Most importantly, all the processes employed in future biorefineries will have to be environmentally friendly. It is essential that we use clean technologies and apply green chemistry principles throughout the biorefinery so as to minimise the environmental footprints of its products and ensure its sustainability (Clark *et al.*, 2006; see Chapter 3).

Biorefineries will be multidisciplinary in nature and would therefore require operators with very different skills and expertise (e.g. agriculturalists, chemists and biotechnologists). The recent formation of new industrial alliances between agribusiness giants, such as Tate & Lyle and Cargill, with well-established chemical companies, such as Dow and Dupont, and upcoming biotechnology industries (including the likes of Genecor and Novozymes) already demonstrates the paramount importance of cross-sector collaborations (Realff and Abbas, 2004).

Products

One of the main drivers for the use of bioenergy and bioproducts is their potential environmental benefits (e.g. carbon dioxide emission reduction, biodegradability). It is thus essential that we assess the environmental impact of all the energy and chemical products we manufacture (across their life cycle) to make sure that they

Table 1.4 General chemical compositions of selected biomass components and petroleum (Pun et al., 2007)

Biomass Component	Chemical Composition	Petroleum Component	Chemical Composition
Cellulose/starch	$[C_6(H_2O)_5]_n$	Gasoline	$\sim C_6H_{14}$-$C_{12}H_{26}$
Hemicellulose	$[C_5(H_2O)_4]_n$	Diesel	$\sim C_{10}H_{22}$-$C_{15}H_{32}$
SW lignin	$[C_{10}H_{12}O_4]_n$		

are truly sustainable and present real (environmental and societal) advantages compared to their petroleum-derived analogues (Gallezot, 2007).

A major issue for biomass as a raw material for industrial product manufacture is variability. Questions of standardisation and specifications will therefore need to be addressed as new biofuels, biomaterials and bioproducts are introduced onto the market. Another major challenge associated with the use of biomass is yield. One approach to improve/modify the properties and/or yield of biomass is to use selective breeding and genetic engineering to develop plant strains that produce greater amounts of desirable feedstocks, chemicals or even compounds that the plant does not naturally produce (Fernando et al., 2006). This essentially transfers part of the biorefining to the plant (see Chapter 2 for some example of oils with modified fatty acid content).

In contrast to fossil resources, biomass feedstocks are composed of highly oxygenated and/or highly functionalised chemicals (see Table 1.4). From an energy point of view, this means that the calorific value of biomass is substantially lower than those of fossil fuels (oxygenated compounds don't burn well!). It is therefore preferable to treat biomass before using it as an alternative fuel or source of energy (see Chapter 6 – Production of Energy From Biomass). This also means that we must apply significantly different chemistries to such highly functionalised chemicals so as to build these up into the valuable chemical products our society is built on (Clark, 2007). In fact, since the production of commodity and speciality functionalised chemicals from fossil resources typically requires highly energy-intensive processes, biomass represents a particularly attractive alternative source of these valuable compounds (Sanders et al., 2005). As highlighted by Ragauskas, the use of carbohydrates as a raw material for chemical production could potentially eliminate the need for several capital-intensive oxidative processes used in the petroleum industry (Ragauskas et al., 2006).

Biorefinery Size

Opinions vary widely on the optimal size of future biorefineries. However, it is our belief that biorefineries will correspond to a combination of large-scale facilities (which can take full advantage of economies of scale and enjoy greater buying power when acquiring feedstocks) and small-scale plants (which can keep transport costs to an absolute bare minimum and take full advantage of available process integration technologies) (European Commission, 2005). Their optimal size, which will obviously depend upon the nature of the feedstock(s) processed, the location of the plant and the technologies employed (not 'one size fits all'!), will correspond to a balance between the increasing cost of transporting pre-treated biomass and the decreasing cost of processing as the size of the biorefinery goes up. Proven full-scale technologies and demonstration biorefinery plants are required before commercial-scale biorefineries of any size can be built.

1.5   Conclusions

Current industrial economies are largely dependent on oil, which provides the basis of most of our energy and chemical feedstocks – in fact, over 90% (by weight) of all organic chemicals are derived from petroleum (Witcoff and Reuben, 1996). However, crude oil reserves are finite and world demand is growing. In the meantime, there is increasing concern over the impact of these traditional manufacturing processes on the environment (i.e. the effect of CO_2 emissions on global warming). In order to maintain the world population in terms of food, fuel and organic chemicals, and tackle global warming, it has been recognised by a number of governments that we need to substantially reduce our dependence on petroleum feedstock by establishing a bio-based economy (van Dam *et al.*, 2005).

For this purpose, long-term strategies that recognise the potential of local renewable resources should be developed. Of paramount importance will be the deployment of biorefineries (of various sizes and shapes) that can convert a variety of biofeedstocks into power, heat, chemicals and other valuable materials, maximising the value of the biomass and minimising waste. These integrated facilities will most likely employ a combination of physical, chemical, biotechnological and thermochemical technologies, which ought to be efficient and follow green chemistry principles so as to minimise the environmental footprints and ensure the sustainability of all products generated (a cradle-to-grave approach). Local pre-treatment of (low bulk density and often wet) biomass will be critical to the development of a sustainable infrastructure capable of working with significant quantities of raw material. Thus, specific attention should be given to the development of these (local) processes. The challenge of the next decade will be to develop demonstration plants, which will require cross-sector collaborations and attract the necessary investors required for the construction of full-scale biorefineries.

References

Andersen, M. and P. Kiel, Integrated Utilisation of Green Biomass in the Green Biorefinery, *Ind. Crops. Prod.*, **11**, 129–137 (2000).

Broins Companies, *The Future is Now for Cellulosic Ethanol*, USDA-ERS Biofuels Modeling Workshop, Washington DC (2007). http://www.farmfoundation.org/projects/documents/BroinCoCellulosic.pdf

Clark, J.H., V. Budarin, F.E.I Deswarte, J.J.E. Hardy, F.M. Kerton, A.J. Hunt, R. Luque, D.J. Macquarrie, K. Milkowski, A. Rodriguez, O.J. Samuel, S.J. Tavener, R.J. White and A.J. Wilson, Green Chemistry and the Biorefinery: A Partnership For a Sustainable Future, *Green Chem.*, **8**, 853–860 (2006).

Clark, J.H., Green Chemistry for the Second Generation Biorefinery – Sustainable Chemical Manufacturing Based on Biomass, *J. Chem. Technol. Biotechnol.*, **82**, 603–609 (2007).

de Jong, E., R. van Ree, R. van Tuil and W. Elbersen, Biorefineries for the Chemical Industry – A Dutch Point of View, in *Biorefineries – Biobased Industrial Processes and Products. Status Quo and Future Directions*, B. Kamm, M. Kamm, and P. Gruber (Eds), Wiley-VCH, (2006).

Ecosytems and Human Well-Being, Island Press, Washington (2007) http://www.millenniumassessment.org/

Deswarte, F.E.I., J.H. Clark, A.J. Wilson, J.J.E. Hardy, R. Marriott, S.P. Chahal, C. Jackson, G. Heslop, M. Birkett, T.J. Bruce and G. Whiteley, Toward an Integrated Straw-Based Biorefinery, *Biofuels, Bioprod. Bioref.*, **1**, 245–254 (2007).

Deswarte, F.E.I., Can Biomass Save the Planet?, *Chem. Rev.*, Accepted (2008).

European Commission, *Biomass – Green Energy for Europe* (2005). http://www.managenergy.net/download/r1270.pdf

European Commission, *Biofuels in the European Union: A Vision for 2030 and Beyond*, (2006). http://ec.europa.eu/research/energy/pdf/biofuels_vision_2030_en.pdf

Fernando, S., S. Adhikari, C. Chandrapal and N. Murali, Biorefineries: Current Status, Challenges, and Future Direction, *Energy and Fuels*, **20**, 1727–1737 (2006).

Gallezot, P., Process Options for Converting Renewable Feedstocks to Bioproducts, *Green Chem.*, **9**, 295–302 (2007).

Gravitis, J., Zero Techniques and Systems – ZETS Strength and Weakness, *J. Cleaner Product.*, **15**, 1190–1197 (2007).

Halasz, L., G. Povoden, and M. Narodoslawsky, Sustainable process synthesis for renewable resources, *Resour, Conserv. Recyc.*, **44**, 293–307 (2005).

Hess, J.R., D.N. Thompson, R.L. Hoskinson, P.G. Shaw and D.R. Grant, Physical Separation of Straw Stem Components to Reduce Silica, *Appl. Biochem. Biotechnol.*, **105–108**, 43–51 (2003).

Agenda 2020, *Integrated Forest Products Biorefineries* brochure, www.agenda2020.org/PDF/IFPB_Brochure.pdf

Kadam, K.L., L.H. Forrest and W.A. Jacobson, Rice Straw as a Lignocellulosic Resource: Collection, Processing, Transportation, and Environmental Aspects, *Biomass and Bioenergy*, **18**, 369–389 (2000).

Kamm, B. and M. Kamm, Principles of Biorefineries, *Appl. Microbiol. Biotechnol.*, **64**, 137–145 (2004).

Kamm, B. and M. Kamm, *Development of Biorefineries by Means of Two Current Projects in North and Middle Europe*, 1st International Biorefinery Workshop US DEO and European Commission, Washington DC (2005).

Koschuh, W., V.H. Thang, S. Krasteva, S. Novalin and K.D. Kulbe, Flux and Retention Behaviour of Nanofiltration and Fine Ultrafiltration Membranes in Filtrating Juice from a Green Biorefinery: A Membrane Sreening, *J. Membr. Sci.*, **261**, 121–128 (2005).

Nilsson, D., SHAM – A Simulation Model For Designing Straw Fuel Delivery Systems. Part 2: Model Applications, *Biomass and Bioenergy*, **16**, 39–50 (1999).

Pun, Y., D. Zhang, P.M. Singh, A.J. Ragauskas, The New Forestry Biofuels Sector, *Biofuels, Bioprod. Bioref.*, DOI 10.1002/bbb.48 (2007).

Ragauskas, A.J., C.K. Williams, B.H. Davison, G. Britovsek, J. Cairney, C.A. Eckert, W.J. Frederick Jr., J.P. Hallett, D.J. Leak, C.L. Liotta, J.R. Mielenz, R. Murphy, R. Templer, T. Tschaplinski, The Path Forward for Biofuels and Biomaterials, *Science*, **311**, 484–489 (2006).

Realff, M.J. and C. Abbas, Industrial Symbiosis – Refining the Biorefinery, *J. Ind. Ecol.*, **7**, 5–9 (2004).

Rupp-Dahlem, C., *Cereal-Based Biorefinery: New Initiatives With the BioHub^{TM} Programme*, Renewable Resources and Biorefineries conference, York (2006).

Sanders, J., E. Scott and H. Mooibroek, *Biorefinery, the Bridge Between Agriculture and Chemistry*, 14th European Biomass Conference, Paris (2005). www.biorefinery.nl/fileadmin/biorefinery/docs/sanders_br_the_bridge_between_agriculture_and_chemistry. pdf

H. Schnitzer, *Agro-based Zero Emissions Systems*, Environmentally Degradable Polymers from Renewable Resources Workshop, Bangkok (2006).

Thang, V.H. and S. Novalin, Green Biorefinery: Separation of Lactic Acid from Grass Silage Juice by Chromatography using Neutral Polymeric Resin, *Bioresource Technol.*, Corrected Proof (2007).

Thorsell, S., F.M. Epplin, R.L. Huhnke and C.M. Taliaferro, Economics of a Coordinated Biorefinery Feedstock Harvest System: Lignocellulosic Biomass Harvest Cost, *Biomass and Bioenergy*, **27**, 327–337 (2004).

USDOE and USDA, *Biomass as Feedstock for a Bioenergy and Bioproducts Industry: The Technical Feasibility of a Billion-Ton Annual Supply* (2005). http://www1.eere. energy.gov/biomass/pdfs/final_billionton_vision_report2.pdf

van Dam, J.E.G., B. de Klerk-Engels, P.C. Struik and R. Rabbinge, Securing Renewable Resource Supplies For Changing Market Demands in a Bio-Based Economy, *Ind. Crops. Prod.*, **21**, 129–144 (2005).

van Dyne, D.L., M.G. Blase and L. Davis Clements, A Strategy for Returning Agriculture and Rural America to Long-Term Full Employment Using Biomass Refineries, in *Perspectives on New Crops and New Uses*, J. Janeck (Ed.), ASHS Press, Alexandria (1999).

Witcoff, H.A. and B.G. Reuben, *Industrial Organic Chemicals*, John Wiley & Sons, Inc., New York (1996).

Wright, M. and R.C. Brown, Comparative Economics of Biorefineries Based on the Biochemical and Thermochemical Platforms, *Biofuels, Bioprod. Bioref.*, **1**, 49–56 (2007a).

Wright, M. and R.C. Brown, Establishing the Optimal Sizes of Different Kinds of Biorefineries, *Biofuels, Bioprod. Bioref.*, **1**, 191–200 (2007b).

2

The Chemical Value of Biomass

David B. Turley

Central Science Laboratory, Sand Hutton, York, UK

2.1 Introduction

From the 1860s, after the sinking of the first oil wells, developed and developing economies of the world have rapidly grown to rely on hydrocarbon fossil fuels to power the economy and deliver the material needs of its citizens. In the main this meant turning away from many plant- and animal-derived materials that had been developed in numerous ways to provide both energy and material resources. Change happened rapidly; within 40 years of the start of large-scale oil extraction, use of plant-derived materials shrank dramatically. For example, use of plant-derived feedstocks in the US dropped to only 16% of demand (Morris, 2006) and much of this represented use of wood pulp for papermaking. Today, plant-derived materials probably represent as little as 5% of industry feedstock inputs, though this varies from sector to sector and is difficult to quantify in detail.

Despite the general move towards use of fossil hydrocarbon feedstocks, some plant-derived materials have continued to provide economic or technical benefits that ensure they remain the preferred source of raw materials for industry. For example, cotton still accounts for 38% of all textile production due to its airflow- and temperature-regulating capabilities, which are difficult or costly to replicate with man-made fibres. Linseed oil remains a key feedstock in surface coating and linoleum flooring applications. Plant oils are still widely used in the oleochemicals sector, where coconut and palm oils are widely used in detergent

Introduction to Chemicals from Biomass Edited by James Clark and Fabien Deswarte
© 2008 John Wiley & Sons, Ltd

and soap formulations. In the medical sector, there is still continued reliance on use of plant feedstocks, such as opiates from *Papaver somniferum*, as sources of pharmaceuticals. Plants also serve as potential sources of new and novel medicinal compounds. For example, there is growing interest in the antioxidant properties of phenolic derivatives found in many plant species.

There are a number of economic, environmental and other factors affecting consumer purchasing and industry response to such pressures, and other legislative developments that are prompting reassessment of the opportunities and properties that plant-derived materials can offer. Pressures include:

• Increasing costs of fossil fuels
• Pressure to reduce volumes of waste going to landfill at the end of product life
• Drive and pressure for increased biodegradability associated with applications in sensitive situations or to promote compostability at the end of product life
• The 'greening' of consumer attitudes and increasing concern over the origin of materials and impacts on the environment that arise from their disposal
• Increasing pressure to curb greenhouse gas emissions
• The desire to reduce use of finite fossil fuels.

Clearly, environmental themes feature strongly in the above issues. However, policy drivers around the world to develop biofuels for transport, to address issues of fuel security and greenhouse gas emissions, have stimulated development of large-scale vegetable oil esterification and ethanol fermentation facilities (from starch and sugar crops). These very large-scale developments offer the potential to consider the added value that can be obtained by exploiting by-product streams arising from such processes, which by their very scale of operation offer opportunities to reduce the costs of production. In addition, the supply chains and facilities developed to supply such industries also offer opportunities for centralised collection and management of plant-derived materials, which again, through economies of scale and colocation of processing operations, can reduce the costs of supply of plant materials. In addition, the ongoing development of thermal biomass conversion technologies designed to improve energy provision from renewable biomass resources also offers the potential to exploit biomass-derived oils as an alternative to fossil oils to deliver an almost unlimited number of chemical products through exploitation of the basic carbon and hydrogen building blocks contained in all plant biomass.

In addition, developments in biotechnology, chemistry, enzyme and catalysis research, engineering, processing and extraction technologies are opening, and will continue to open up, new routes, pathways and possibilities to exploit plants in an ever-widening fashion and at lower cost to industry.

This chapter reviews how use of plant biomass for delivery of raw materials for the chemical industry has developed, identifying how key plant metabolites have or are being used by industry, and how the potential to exploit plants can be expanded through use of biotechnology and developing thermal technologies to deliver a wide range of material building blocks for bulk chemical industries, as well as speciality, high-value chemicals for healthcare and niche market outlets.

2.1.1 Key Routes of Plant Exploitation for Chemical Raw Materials

Plants can be exploited in a number of different ways to deliver valuable feedstock materials for industry (Figure 2.1).

Traditionally, non-food uses for crops have made use of particular plant tissues for extraction of specific materials, including:

- Stems for fibres and sugar
- Seeds and tubers for starch
- Roots for sugar and starch
- Seeds, fruit and nuts for oil
- Foliage/stems for essential oils
- Sap for resins and latex
- Various plant parts for medicinal uses and as sources of dyes, resins and waxes.

In many cases, food crops are exploited to provide abundant sources of carbohydrates and oils that are then diverted to industrial uses, for example corn, potatoes and wheat for starch, and oilseed rape and sunflower for oil. In other cases, nonedible crops are commercialised primarily for specific industrial or medicinal use, where examples include linseed, castor bean and rubber palm. Where plants are commercialised for industrial uses, unless products command a very high value, candidate plants must be capable of producing large quantities of the metabolites of interest. In the case of oils for specific uses, this means plants must be enriched in specific fatty acids. This arises from the fact that plant-derived chemicals in many cases compete with petrochemical-derived alternatives and this requires that costs of extraction and refining are kept as low as possible in order to remain commercially competitive.

In the following sections, current and potential future means by which plants can be exploited to produce feedstocks for the chemical industry are reviewed by key plant metabolite sectors:

- Oils
- Carbohydrates – sugars, starch, cellulose, hemi-cellulose
- Lignin
- Proteins
- Waxes
- Secondary metabolites.

The above is followed by an overview of the assessment of the potential for future exploitation of plants for chemical feedstocks through developments in biotechnology and biorefining.

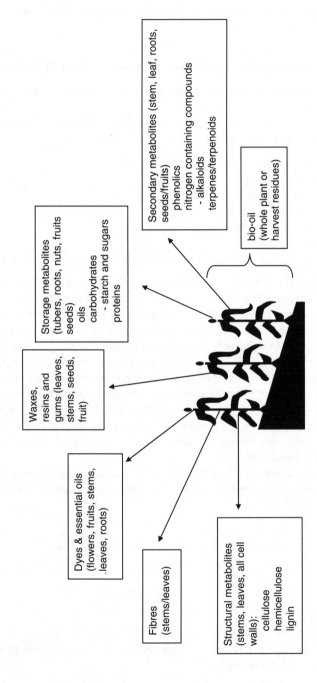

Secondary metabolites (stem, leaf, roots, seeds/fruits)
phenolics
nitrogen containing compounds
- alkaloids
terpenes/terpenoids

bio-oil
(whole plant or harvest residues)

Storage metabolites
(tubers, roots, nuts, fruits seeds)
oils
carbohydrates
- starch and sugars
proteins

Waxes,
resins and
gums (leaves,
stems, seeds,
fruit)

Dyes & essential oils
(flowers, fruits, stems,
leaves, roots)

Fibres
(stems/leaves)

Structural metabolites
(stems, leaves, all cell
walls):
cellulose
hemicellulose
lignin

Figure 2.1 Key sources of plant-derived chemicals and materials for industrial exploitation

2.2   Plant Oils

2.2.1   Abundance and Sources

A wide range of oil-bearing crops grown for food uses are also exploited for non-food uses, but it is difficult to estimate how much is diverted to such uses. Table 2.1 illustrates the scales of world production for the most abundant and commonly traded oil crops that also have well-developed industrial uses.

The tropical oil crops, coconut and palm, are the most efficient oil-producing crops, with coconut plantations yielding up to 2 tonnes per hectare of oil and the best performing palm plantations from 5–6 tonnes per hectare. By comparison, oil yields of temperate oil crops are typically of the order of 1–2 tonnes per hectare for the best oil-yielding crops (oilseed rape and sunflower). Clearly, Table 2.1 represents only a small fraction of oil-bearing plant species. Many other seed, fruit and nut oils are extracted for food use, however unless they contain fatty acid profiles or fatty acid derivatives of specific industrial interest, total oil-yield, fatty acid yield and cost of the final oil product tends to limit their use in industrial applications on all but a small or localised scale.

Many wild and currently non-commercialised plant species contain oils that may potentially have industrial applications, though a full description of profiles for all species is beyond the scope of this current chapter. Those interested in the wider potential are referred to Padley *et al.* (1994) (The Lipid Handbook), which provides a detailed list of major and minor oil-bearing plants, seeds and nuts, and their typical oil profiles. Such reviews demonstrate the variability that can occur in fatty oil profiles within closely related species. For example, the genus *Cuphea* represents a wide range of closely related species whose potential is being examined for exploitation as a temperate source of fatty acids with carbon chain lengths of C10 to C14 for use in the soap and surfactant sector. *Cuphea* species possess a wide range of fatty acid profiles that differ significantly between species and even sub-species, which highlights one of the problems associated

Table 2.1   *World production (2006) of major oilseed crops (FAOSTAT) and typical oil contents for oil crops widely used by industry*

Oil plant	Area harvested (million hectares)	Seed production (million tonnes)	Typical oil content (%)
Castor	1.26	1.40	44–45
Coconut (copra[a])	9.43	4.95	65–68 (as copra)
Linseed	3.01	2.57	39–42
Palm kernel	9.84	9.66	21–25
Palm fruit			46–57
Soy bean	92.98	221.50	16–18
Sunflower	23.78	31.32	40–50

[a] Copra is the dried endosperm of the coconut from which oil is extracted.

with prospecting for promising candidate species, and in ensuring that research programmes are utilising similar plant material.

In recent decades, vegetable oil production and use has been growing at around 4% per annum. The internationally focused Food and Agriculture Organisation estimates that around one third of this increase results from the growing industrial use of vegetable oils. The increasing interest in use of vegetable oils as feedstocks for production of biodiesel, driven by environmental and energy security concerns, will in the short and medium term significantly increase the volumes of vegetable oil destined for non-food uses.

2.2.2   Oil Profiles of Major Oil Crops

The most concentrated reserves of oils in plants are found in fruit (e.g. palm kernel oil), nuts and seeds. Typically, triacylglycerides (often referred to as triglycerides) make up 93–98% by weight of plant-derived oils. These fatty-acid esters of glycerol comprise a glycerol molecule with typically, though not exclusively, three different fatty acids (see Figure 2.2). The prominence of individual fatty acids in individual triacylglycerides can be influenced by plant breeding methods. Commercialised examples of this include high erucic acid cultivars of oilseed rape and the development of 'high oleic' cultivars of rape and sunflower for food and industrial markets (discussed later).

All vegetable and plant oils comprise a mixture of different triacyglyceride molecules, which differ between plant species, sometime subtly and in other cases dramatically, in the ratio of different types of fatty acyl chains that they contain. The dominance of different fatty-acid carbon chain lengths and the degree of saturation of individual fatty acids within an oil impart different physical and chemical characteristics to oils from different plant species. Long-chain fatty acids (C18) dominate in temperate commercial oil crops, while shorter fatty acid chains (C12–C14) dominate tropical commercial oil crops like coconut and palm. The degree of saturation of fatty acids affects their reactivity and resistance to oxidative degradation (which is faster with less saturation), while chain length affects melting point and influences characteristics such as viscosity at room temperature, which

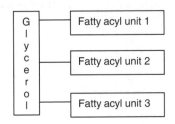

Figure 2.2   *Schematic representation of a triacylglycerol molecule*

can significantly affect use where oils are used in bulk in relatively unmodified form. Table 2.2 illustrates the typical fatty acid profiles for a range of oils widely used for industrial purposes.

Mineral oil-derived fatty acids are shorter and more branched than vegetable-derived fatty acids and are always saturated and much more resistant to biodegradation. This suits use in high-temperature or high-wear situations, such as in engines. However, vegetable-based oils still find many uses in industry.

- *Lubricants:* The base oil constitutes 90–95% by weight of lubricant formulations, with the physical properties of the oil reflecting the properties of the individual fatty acids in the formulation. The property of increased biodegradability associated with vegetable oils is a positive benefit for lubricants used in sensitive marine or forest environments. To increase the working life and temperature range of vegetable-based lubricants, antioxidants and 'pour-point depressant' additives are used.
- *Surfactants:* The hydrophobic chain and hydrophilic 'head' of fatty acid molecules makes them well suited for use in surfactant formulations in detergent and cleaning formulations.
- *Soaps:* Short-chain saturated fatty acids, particularly C12 (lauric) from coconut, are used in cosmetic soap manufacture. Though there has been interest in developing Cuphea species as a source of C12 fatty acids in temperate northern latitudes, it has proved difficult to commercialise to date.
- *Surface Coatings:* The paint industry traditionally uses vegetable-derived long-chain fatty acids with a degree of unsaturation. Linseed oil, tall oil (a by-product from wood pulping for paper making) and soy oil are used in alkyd gloss paints, though high levels of linseed oil can lead to yellowing (through oxidation of its high linolenic acid content).
- *Solvents:* Fatty-acid esters have good solvating properties and are used as cleaning agents in metal-working and printing industries (Turley *et al.*, 2004).
- *Polymers:* Unsaturated fatty-acid chains offer opportunities for polymerisation that can be exploited to develop uses in surface coatings and plastics manufacturing. Polyunsaturated fatty acids can be dimerised to produce feedstocks for polyamide resin (nylon) production. Work is also ongoing to develop polyurethanes from vegetable oils through manipulation of functionality in the fatty-acid chains, to produce both rigid foams and elastomers with applications in seals, adhesives and moulded flexible parts (see Chapter 5 for more information).
- *Plasticisers and Slip Agents:* Fatty acids can also be incorporated into fossil-derived plastics as plasticisers or lubricants. For example erucamide, commercially derived from high-erucic acid oilseed rape and Crambe (Crambe abyssinica), is used as a surface lubricating slip agent in the plastics industry. The fatty-acid amide migrates to the surface of polyurethane and other plastics and acts as a lubricating surface during heat extrusion operations.

Table 2.2 Fatty-acid ranges and composition (% by weight) for a range of common crop-derived oils that have non-food uses (derived from Padley et al., 1994 and author's own data for linseed)

Fatty Acid	Degree of Saturation[a]	Common Name	Rapeseed (low erucic)	Sunflower	Linseed	Soy	Coconut	Palm
C6:0	S	Caproic					0–0.6	
C8:0	S	Caprylic					4.6–9.4	
C10:0	S	Capric					5.5–7.8	
C12:0	S	Lauric		0–0.1		0–0.1	45.1–50.3	0–0.4
C14:0	S	Myristic	0–0.2	0–0.2		0–0.2	16.8–20.6	0.5–2.0
C16:0	S	Palmitic	3.3–6.0	5.6–7.6	6.0	8.0–13.3	7.7–10.2	40.1–47.5
C16:1	M	Palmitoleic	0.1–0.6	0–0.3		0–0.2		0–0.6
C17:0	S	Margaric	0–0.3					
C17:1	M	Heptadecenoic	0–0.3					
C18:0	S	Stearic	1.1–2.5	2.7–6.5	2.5	2.4–5.4	2.3–3.5	3.5–6
C18:1	M	Oleic	52.0–66.9	14.0–39.4	19.0	17.7–26.1	5.4–8.1	36.0–44.0
C18:2	P	Linoleic	16.1–24.8	48.3–74.0	24.1	49.8–57.1	1.0–2.1	6.5–12.0
C18:3	P	Linolenic	6.4–14.1	0–0.2	47.4	5.5–9.5	0–0.2	0–0.5
C20:0	S	Arachidic/Arachic	0.2–0.8	0.2–0.4	0.5	0.1–0.6	0–0.2	0–1.0
C20:1	M	Eicosenic/Gadoleic	0.1–3.4	0–0.2		0–0.3	0–0.2	
C20:2	P	Eicosadienoic	0–0.1			0–0.1		
C22:0	S	Behenic	0–0.5	0.5–1.3		0.3–0.7		
C22:1	M	Erucic	0–2	0–0.2		0–0.3		
C22:2	P	Brassic	0–0.1	0–0.3				
C24:0	S	Lignoceric	0–0.2	0.2–0.3		0–0.4		
C24:1	M	Nervonic	0–0.4					

[a] S = saturated, M = monounsaturated, P = polyunsaturated

2.2.3 Oils With Modified Fatty-Acid Content

To exploit additional market opportunities or to improve oil characteristics, traditional plant breeding methods have been used to create oils with modified fatty acid profiles.

High Oleic Oils

Oils with a high degree of unsaturation are susceptible to oxidative and thermal breakdown, a disadvantage for use in industrial lubricant applications, particularly in high wear situations. Increasing saturation levels in oils also increases their plasticity. Through conventional breeding, sunflower cultivars with increased levels of oleic and oleic plus palmitic acid content have been developed (Table 2.3). Sunflower oils with as high as 90% oleic acid content have been developed. High oleic hybrids of oilseed rape, soy and safflower (*Carthamus tinctorius*) are also available.

2.2.4 High Erucic Acid Oils

Though early oilseed rape (*Brassica napus*) breeding programs were developed to select for low erucic acid content in oils for the food sector (in favour of increased oleic content), high erucic acid oilseed rape (HEAR) cultivars were developed for the non-food sector (oils containing up to 50% erucic acid – see Table 2.4). This is the base feedstock for production of erucamide, erucyl alcohol, erucate wax esters, and pelargonic and brassilic acids through various chemical transformations. These materials have important applications as slip agents, in printing ink formulations, as detergents and coating agents, and in polyesters and nylons.

Table 2.3 *Fatty acid composition (%) of sunflower oils from conventional, high oleic and high oleic plus high palmitic cultivars (Guinda et al., 2003)*

			Cultivar type		
Fatty Acid	Degree of Saturation[a]	Common Name	Conventional	High oleic	High oleic & High palmitic
C14:0	S	Myristic			1.2
C16:0	S	Palmitic	6.4	4.7	38.7
C16:1	M	Palmitoleic			0.2
C18:0	S	Stearic	5.0	3.8	4.5
C18:1	M	Oleic	29.3	80.2	42.1
C18:2	P	Linoleic	58.3	9.5	12.3
C18:3	P	Linolenic		0.3	0.3
C20:0	S	Arachidic/Arachic	0.3	0.4	0.5
C20:1	M	Eicosenic			0.2
C22:0	S	Behenic	0.7	1.1	

[a] S = saturated, M = monounsaturated, P = polyunsaturated

Table 2.4   Fatty acid profile (% by weight) of oil from conventional and high erucic acid rape (HEAR) cultivars (Padley et al., 1994)

Fatty Acid	Degree of Saturation[a]	Common Name	Conventional (low erucic)	High erucic
C16:0	S	Palmitic	4	3
C16:1	M	Palmitoleic		
C18:0	S	Stearic	2	1
C18:1	M	Oleic	56	16
C18:2	P	Linoleic	26	14
C18:3	P	Linolenic	10	10
C20:0	S	Arachidic/Arachic		1
C20:1	M	Eicosenic	2	6
C22:1	M	Erucic		49

[a] S = saturated, M = monounsaturated, P = polyunsaturated

These two examples clearly demonstrate the potential for development and manipulation of fatty acid profiles to improve plant oil characteristics for industrial use, and in the case of HEAR, to improve the economics of processing, reducing costs to the end user. However, one of the downsides of manipulating fatty-acid profiles in crops that have both food and non-food uses is the need to keep crops separated on the farm and in the post-farm-gate supply chain, to avoid risks of contamination of either chain.

2.2.5   Novel Fatty-Acid Derivatives Found in Plants That Have Industrial Uses

There are a number of unsaturated derivatives of plant-derived fatty acids that have, or could have, interesting industrial applications:

- Epoxidised oils (with isolated double bonds), such as epoxidised soybean are used as plasticisers. The natural occurrence of epoxidised fatty acids could be exploited to reduce processing costs on the pathway to polymerisation. For example, seed oils derived from *Vernonia galamensis* and *Euphorbia lagascae* both contain significant proportions (60–80%) of the epoxy acid vernolic acid, with interesting applications in greases and polyurethane applications (Turley *et al.*, 2000).
- Petroselinic acid, an isomer of oleic acid, is found in many seed oils of the *Umbelliferae* family (ranging from 50–90% of oil composition). It can be oxidised to adipic and lauric acids, and may have pharmaceutical and cosmetic applications. Coriander is being evaluated as a potential source of this fatty acid in Europe.
- The hydroxy fatty acid, ricinoleic acid (C18:1 (OH)) can account for up to 90% of castor bean oil composition. It is mainly used as an industrial oil in paints and varnishes. It is also used as a plasticiser in soaps, waxes, polishes and

hydraulic fluids, and as a lubricant (particularly after hydrogenation). It is also a key feedstock for a number of derivatives used in the perfume, nylon and textile industries. Plants of the *Lesquerella* genus also contain hydroxy fatty acids of interest, and work to commercialise this species has started in the US to provide a domestic source of feedstock materials.

2.2.6 Industrial Uses for Glycerol

Saponification, hydrolysis and esterification of vegetable oils to release fatty acids for the oleochemical uses detailed above, results in production of glycerol as a by-product. Glycerol is an important platform molecule in its own right, with numerous industrial uses (see Chapter 6 for more information).

Traditionally, glycerol is used in the pharmaceutical and personal-care sectors and to produce polyethers for polyol foams, alkyd resins (used in ink formulations) and ethylene glycol (anti-freeze). However, many of these markets are saturated or declining in the face of competition. As glycerol is a relatively expensive polyol, price affects its uptake and use. As increasing volumes of biodiesel are produced from vegetable oils via large-scale transesterification operations, increasing tonnages of glycerol are becoming available, as 100 kg of glycerol is produced for every tonne of biodiesel produced. This increased availability and associated falling price could make it an increasingly attractive material for industry, stimulating interest in further chemical transformation.

New and novel uses for glycerol are being developed including:

* Use in glycerol-based surfactants for degreasing, perfume and printing inks and as a non-volatile reactive solvent in paints
* Use as a feedstock for production of epichlorohydrin (via dichloropropanol) for use in production of epoxy resins used in paper reinforcement and water purification treatments
* Development of polyglycerol ester polymers, for use as surfactants and lubricants
* Production of highly branched glycerol-based polymers including polylactide polymers
* Catalytic reduction and/or fermentation to produce 1,3-propanediol, a polymer with applications in the textile sector and a key feedstock for production of the renewable polymer 'Sorona' produced by Du Pont
* Use of di- and tri-ethylene glycols as gas scrubbers.

These developments offer significant potential to expand the market demand for glycerol.

2.3 Carbohydrates

Carbohydrates account for the greatest proportion of plant biomass, with cellulose and hemi-cellulose acting as major structural components of plant tissues and plant cell walls. Starch held in seeds and tubers, and sugars (sucrose) held in plant stems or roots act as storage reserves for future growth and development. As readily available and relatively cheap sources of material, plant carbohydrates have been widely exploited for industrial uses. In terms of simple sugars, the main sources of commercial production are sugar cane in tropical climates and sugar beet in temperate climates.

2.3.1   Starches and Sugars

Outside of the use of cellulose for papermaking, starch is the most widely used plant-derived carbohydrate for non-food uses. Around 60 million tonnes of raw starch are produced per year for food and non-food uses. The US accounts for most of the world's production, utilising starch from maize, which accounts for over 80% of world production. The starch market in the US is driven by the large isoglucose sweetener market and now increasingly by the growing bioethanol market, which uses maize as a fermentation feedstock. Europe derives most of its starch from wheat and potatoes, which account for 8% and 5% of world starch production, respectively. The other main source of starch is cassava (tapioca), produced in South East Asia. Small amounts of oat, barley and rice are also exploited for starch production. Many edible beans are also rich in starches, but are not commonly exploited for non-food uses.

Starch production in Europe is currently constrained by a starch quota system that offers financial incentives to first processors, designed to prevent over-production in the food sector. Opening up of new industrial market outlets would enable a reassessment of current quota limits imposed on member states, and enable wider participation in starch production across the European Union.

Sales of bran and gluten, produced as by-products of maize and wheat starch production, help keep the cost of maize and cereal starch lower than that of potato starch. However, starches derived from different plant sources are not necessarily directly interchangeable, due to the different compositional traits of different starches and the specific requirements of the end user. Plant-derived starch is typically made up of 10–20% of the straight chain α-glucose polymer amylose and 80–90% of the branched α-glucose polymer amylopectin. Starches with high amylose contents are favoured in film, foam and adhesive applications, while high amylopectin starches (so-called waxy starches) are preferred for surface-coating applications (e.g. in paint formulations). Starch from the Quinoa plant has a naturally low amylose content and has been evaluated for use in papermaking and the adhesive sector.

Starches may be used directly as feedstocks, or in more technical uses in modified form (typically as starch esters and ethers), or simply converted to glucose syrups for use in industrial fermentation processes or for onward conversion to isoglucose (fructose). After use of unmodified starches in ethanol production, the largest industrial user of both unmodified and modified starches is the papermaking industry. Starches bind with cellulose fibres to add strength to the paper during the paper production process. They are also applied to finished papers as a surface coating to reduce water infiltration, further strengthen the paper and improve printing properties. Modified starches are also used in glues used in corrugated cardboard manufacture. Large volumes of starch are also used in plasterboard materials produced for the construction industry. Other direct uses of starch include:

- Finishing agents for yarns and textiles
- Fillers in biodegradable plastics
- Cationic starches to coagulate or flocculate materials during wastewater treatment
- Detergent formulations
- Packaging materials (expanded polymer foams)
- Binders and coating agents in agrochemical and pharmaceutical applications
- Stabilisers in gloss and emulsion paints as a replacement for acrylic monomers.

Relatively unmodified starches have made good in-roads into the biodegradable packaging market, both as a loose-fill expanded (blown) starch product and as moulded forms for use in applications such as fast-food tableware. In addition, by mixing starch with small quantities of biodegradable synthetic polymers (polyvinyl alcohol or polycaprolactone) biodegradable films have been developed that are suited to applications with limited technical demands, including biodegradable bags, wrapping and disposable packaging. Their mechanical properties typically lie between those of low- and high-density polypropylene.

One of the anticipated growth areas for industrial uses of plants is in development of non-brittle, durable polymers from renewable plant feedstocks (in both biodegradable and non-biodegradable forms). Starch and sugars are currently used commercially as feedstocks for polyester production utilising microbial monomer and polymer fermentation systems (see Chapter 5 for more information). Current commercialised examples include:

- Polyhydroxybutryate (PHB)
- PHB plus polyhydroxyalkanoates (PHA)
- Poly-β-hydroxy butyrate-co-valerate (PHBV)
- Polylactic acid (PLA).

Other fermentation polymer products in development include polyhydroxybutyrate-co-polyhydroxyhexanoates (PHBH).

PLA is finding a wide range of uses, both as a moulded product for the packaging sector, as a film for wrapping and lamination applications, as a fibre fill (pillows/duvets etc.), as a foam, and as a spun textile fibre. Lactic acid produced from microbial fermentation of starch also has many other industrial opportunities, as it is a useful base chemical feedstock for a range of applications. Lactic acid can be chemically converted to propylene glycol, the base for a range of polymer derivatives, to propylene oxide to produce epoxides, to acrylic acid to produce polyacrylic acid and resins, and to lactate esters, amongst other possibilities (Vink, 2002). The ethyl lactate derivative is also used as a solvent in the electronics industry.

There is also US research interest in using pectin in polymer applications. Pectin is a complex plant cell wall heteropolysaccharide (based on galactose, rhamnose, arabinose and xylose) that can be blended with synthetic polyvinyl alcohol (PVA) to produce biodegradable polymers with a wider range of properties than those of starch-based polymers alone. The new pectin/PVA biodegradable polymer should be capable of replacing conventional PVA applications in blow-moulded, extruded, film and injection-moulded applications.

The potential for fermentation technologies to produce bulk quantities of chemicals from sugars has stimulated much interest. The EU BREW project (BREW, 2006) identified a significant number of potential chemical building blocks of current commercial interest, which are, or could, be economically produced from fermentation and/or enzymic conversion of glucose, sucrose or starch (Table 2.5).

2.3.2   Cellulose

Cellulose is another ubiquitous carbohydrate in the plant kingdom. Cellulose is a linear polymer of β-glucose, unbranched and resembling amylose, which has been used for a number of industrial applications. Around 95% of cellulose production is used in papermaking, derived from wood-pulping operations. Wood contains up to 40–50% cellulose by mass. A number of important textile fibres are derived from cellulose including Viscose (cellulose xanthate), which accounts for the greatest use of cellulose after papermaking. Cellulose acetate is an industrially important ester of cellulose and is used in a wide range of products, including cigarette filters, films and other coating materials. It is also used in moulded articles (e.g. spectacle frames), as a micro- and ultra-micro-filtration membrane in the pharmaceuticals sector and as a matrix for administering slow-release steroid drugs.

Cellulose ethers are a wide-ranging family of cellulose derivatives, commonly used in the food and pharmaceutical industry. Methyl, ethyl and propyl cellulose esters are used as drilling aids in mining and as detergents. They are also used as coatings and adhesives in cosmetic and pharmaceutical products.

Table 2.5   *Key plant carbohydrate–derived chemical building blocks identified in the EU Brew project*

Acetic acid	Fumaric acid	Malic acid
Acetone	Glutamic acid	Propionic acid
Butanol	Gluconic acid	Succinic acid
Citric acid	Itaconic acid	
Ethanol	Lactic acid	

2.3.3   Hemicellulose

Hemicellulose is a key component of plant cell walls, comprising up to 25–30% of woody plant tissues. It is a branched polymer that may contain many different sugar monomers, hexoses and pentoses, though xylose is always present in the largest quantity. The sugar monomers include glucose, mannose, galactose, rhamnose, and arabinose. Oxides of sugars may also be present as acids e.g. mannuronic acid and galacturonic acid. In contrast to cellulose, hemicellulose is easily hydrolysed into its constituent monomers.

Hemicelluloses can be hydrolysed into their component sugars and used as a fermentation feedstock for the production of ethanol and other alcohols (e.g. butanol, arabitol, glycol and xylitol), organic acids (e.g. acetic acid), acetone and gases (e.g. methane and hydrogen). The wider monosaccharide profile offers opportunities to develop different products to those derived from glucose alone.

Pentoses contained in hemicellulose are used to produce furfural, a useful industrial chemical, used as a solvent for resins and waxes and in petrochemical refining. It is also used as a feedstock for a range of aromatic substances (it has an almond odour) including preservatives, disinfectants and herbicides. Furfural can be converted to furfuryl alcohol and used to make resins for composite applications with fibreglass and other fibres. These are of interest in the aircraft component and automotive brake sectors. Furfural is commercially derived from acid hydrolysis of waste agricultural by-products, such as sugarcane bagasse, corn cobs and cereal brans. Around 450 000 tonnes is produced by this method per year.

2.4   Lignin

Lignin is the third most abundant structural polymeric material found in plant cell walls typically comprising up to 20–30% of woody biomass, from which most lignin is sourced as a by-product of papermaking. Lignin binds hemicellulose and cellulose together in plant cell walls and shields them from enzymic and chemical degradation.

Lignin is the main non-carbohydrate polymer found in plants. It is a complex, highly aromatic, polyphenolic material with a complex, cross-linked structure derived principally from coniferyl alcohol.

Lignin is used in many industries. Unsulfonated lignins are used in the production of vanillin and dimethyl sulfoxide. Around 1.4 million tonnes of unsulfonated lignins are produced per annum and used in the polymer resin industry to manufacture plywood. Alkali treated lignins could also be a potential feedstock for polyurethane industries and used in the production of novel polyurethane composites.

Sulfonated lignin is predominately used as a stabiliser in drilling muds and emulsions. However it has a wide range of other potential, though relatively low value, applications, including use as a dispersant in paints, clay, porcelain, dyes, pesticides and industrial cleaning agents. It is also used as a binder and filler in the pelletisation of animal feed, as a substitute for carbon black in tyres and as an additive to concrete and gypsum.

2.5 Proteins

Seeds of legumes, oilseed meals (the material left after oil extraction) and, to a lesser extent, wheat grains are important sources of protein, most commonly exploited commercially as animal feed supplements. In de-oiled oilseed meal, protein content can reach 40%. There are many potential cosmetic and technical uses for such vegetable proteins. However, extraction of high quantities of pure protein compounds is difficult without affecting their structural integrity or chemical properties. Development of methods to improve protein extraction is the subject of current research and development. To date most functional plant-derived proteins have been crude extracts from soya and wheat, primarily used in the food sector. However, there are opportunities to use proteins in paper-coating treatments, adhesives (hot melt using wheat gluten) and in the cosmetic sectors (haircare, skincare, bathing products). Work in the EU Enhance project, led by INRA in France, has led to the isolation and purification of protein fractions from oilseed rape meal capable of substituting for both soya and casein proteins. Protein composition is affected by plant genotype. For example, breeding developments in oilseed rape associated with moving from 'old' cultivars to current 'double-low' cultivars (low in erucic acid and glucosinolates) has decreased total protein content, but increased cruciferin at the expense of napin protein constituents (Enhance, 2003). The protein fractions themselves can be modified by processing, for example to increase solubility, to optimise emulsifying and foaming capabilities, or film-forming capabilities to increase opportunities for use in biopolymer markets.

2.5.1 Healthcare Proteins

In addition to the above, there is interest in the development of specific proteins in plants for the healthcare sector, such as antibodies, antigens and vaccines, by utilising transgene technologies (see Table 2.6). This technology offers the potential for significant and more rapid scale-up of production and at relatively low cost compared to current fermentation approaches. Also, more complex proteins, such as anticlonal antibodies, which are difficult to produce in current fermentation systems, can be produced in plant systems. Some pharmaceutical proteins can only be produced in plants, such as secretory IgA antibodies and recombinant immune complexes (Ma, 2007). Current levels of expression are relatively low, typically around 1% of total soluble protein content. However, conventional breeding techniques offer opportunities to significantly enhance yields. Demands for use of such materials differ, but the potential for very high returns associated with such markets is likely to be available to only a small number of specialist growers.

Table 2.6   *Current plant-derived therapeutic proteins on the market or near commercialisation*

Target trait	Target crop
β–glucuronidase trypsin (commercialised)	Maize
Lactoferrin (commercialised)	Rice
Collagen	Tobacco
Antibodies for tooth decay and non-Hodginsons lymphoma	Tobacco
Human gastric lipase	Maize
Therapeutic enzymes and dietary supplements	Maize
Hepatitis B and Norwalk vaccines	Potato
Rabies vaccines	Spinach
Insulin	Safflower

2.6   Waxes

Plant waxes are concentrated on leaves and leaf sheaths and on fruit skins, or in some exceptional cases in the seeds of plants. Most vegetable waxes contain predominantly wax esters plus a variety of other lipid materials, which affect the degree of saturation and other properties of the wax derived from different sources. Most use has been made of plant waxes in the cosmetic sector, but there is increasing interest in the use of plant-derived sterols as dietary supplements to reduce cholesterol formation.

The jojoba plant (*Simmondsia chinensis*) produces wax esters rather than tri-acylglycerols in its seeds and it has become a significant crop for the cosmetic sector. Its wax consists mainly of long-chain fatty acids linked to long-chain fatty alcohols.

The leaves of the carnauba wax palm, *Copernica prunifera*, have a thick coating of wax, which can be harvested from the dried leaves. It contains mainly wax esters (85%), accompanied by small quantities of free acids and alcohols, hydrocarbons and resins.

2.7   Secondary Metabolites

Having accounted for storage carbohydrates, triglyceride oils, structural metabolites and waxes, much of what remains is commonly termed 'secondary metabolites'. Across the plant kingdom this represents a vast array of chemicals typified by flavanoids, terpenes, phenols, alkaloids, sterols, tannins, sugars, gums, suberins, resin acids, carotenoids and many others. Secondary metabolites in many cases represent materials that help to prevent herbivory and fungal attack in plants, as well as being involved in signalling between plants, helping to ward off competitors through processes such as allelopathy and in attracting other species such as pollinators and species involved in seed dispersal, through scents, flavour and pigment molecules. Other metabolites may be involved in helping plants cope with environmental stress. For example, flavenoids help protect plant foliage from the damaging effects of UV radiation.

Individual chemicals can be restricted to certain species and tissues and in some cases to certain cells – such as trichomes, oil-producing glands found on the leaves of some plants.

Secondary metabolites include essential oils, used in the flavour and fragrance industries. Essential oils are found in over 50 plant families and represent terpenoids and other aromatic compounds accumulating typically at relatively low concentrations (usually <1% of fresh weight, but can be up to 20%), but which have useful antimicrobial activity (Biavati *et al.*, 2003). Production of essential oils by plants is affected by many factors influencing plant growth.

Secondary metabolites also include many plant-derived chemicals used in pharmacy applications or as non-prescription health supplements. Species that are still directly exploited for pharmaceutical use are listed in Table 2.7.

In a recent review (Fowler, 2006) it was identified that plant-derived drugs are still an important part of the pharmaceutical armoury in a plant-derived pharmaceutical market worth over $20 billion per year. However, while plants are responsible for a very significant proportion of drug discoveries, there are only a few examples where plant components and extracts are used directly (accounting for 6% of all prescribed drugs), these include codeine, morphine, taxol and cannabis (in development for treatment of multiple sclerosis patients). In other cases, semisynthetic formulations (accounting for 27% of all prescribed drugs) are used to enhance effectiveness, while synthetic mimics of originally plant-derived active molecules account for around 23% of natural product or natural product inspired drugs, where examples include salicylic acid, the active ingredient of aspirin, derived originally from willow (*Salix* sp.) bark. However, the potential for further exploitation of plant chemicals is significant (Fowler, 2006). New tools to help prospect plant chemical libraries could also help speed up the process of identifying new potential actives and drug leads.

Table 2.7   *Important plant-derived pharmaceuticals*

Drug	Activity	Key plant source
Terpines		
Artemisinin	Antimalarial	*Artemisia annua*
Paclitaxel/Taxol	Anticancer	*Taxus brevifolia*
Alkaloids		
Codeine, Morphine	Analgesics	*Papaver somniferum*
Cocaine	Analgesic	*Erythroxylum coca*
Galanthamine	Cholinesterase inhibitor	*Leucojum aestivum*
Quinine	Antimalarial	*Cinchona ledgeriana*
Vinblastin/Vincrystine	Anticancer	*Catharanthus roseus*
Steroids		
Digoxin	Cardio-vascular moderator	*Digitalis lanata*

Table 2.8 Examples of polyphenolic bioactive compounds found in plants

Compounds	Examples	Activity
Flavonols	Quercetin, kaempferol	Reduced risk of chronic disease
	Catechin, epichatechin	Increased antioxidant activity
	Procyanidin	Increased antioxidant activity, decreased LDL[a]
Flavanones	Naringenin, hesperidin	Decreased lipids
Isoflavones	Genistein, daidzein	Estrogenic activity
Stilbenes		Inhibits LDL oxidation, anticancer
Stanols and sterols	Sitostanol, sitosterol, campesterol, campestanol	Reduced cholesterol absorption

[a] LDL = low density lipoprotein

With growing research into the impact of diet on health and well-being, there has been a growing interest in the impact of plant-derived bioactive compounds on human health, though in many cases there is limited clinical evidence to support the claims made. However, around 100 flavenoid-containing medicines are available in countries, such as France, that permit their use as alternative or complimentary therapeutics, primarily as treatments for the vascular system. Phenolic derivatives from plants have been found to exert beneficial impacts in reducing blood cholesterol levels, decreasing absorption of lipid materials and acting as antioxidants. Table 2.8 lists a few key polyphenol compounds of interest (drawing on work by Hooper and Cassidy, 2006).

2.7.1 Glucosinolates

The Brassica family produces a wide range of glucosinolate compounds, anionic glycosides produced by the plant as antifeedant protective chemicals. As significant amounts of these compounds are left in the oilseed rape meals that remain after oil extraction, there is currently interest in exploiting these materials as crop-protection products for control of soil-borne diseases (Palmieri, 2003).

2.7.2 Other Industrial Uses for Secondary Metabolites

Oleoresin derivatives from plants such as turpentine and rosin, and their associated tars, and pitches have been obtained from wood tapping and wood processing operations for centuries. Turpentine is still widely used as a solvent in paints and varnishes. It can be fractionated into its component parts, but this is only economically viable on a large scale. The monoterpene α-pinene is used to prepare synthetic pine oil (the biggest single turpentine derivative), polyterpine resins, perfumes/fragrances and insecticides. In the US, 90% of pine oil is used in cleaning and disinfectant formulations and around 330 000 tonnes of turpentine are produced per annum from wood sources. A range of low molecular-weight polymers can be produced from tree-derived monoterpenes and diterpines by cationic polymerisation. These polymers are used to impart 'tack' in compounds for pressure-sensitive and hot-melt adhesives.

2.8   Prospects Arising from Developments in Plant Biotechnology and Biorefining

The preceding sections highlight how specific plant metabolites can be utilised for industrial uses on both large and small scales. However, low yields of target metabolites, variation in quality or composition of metabolites, presence of contaminants or undesirable characters and additional costs of processing and extraction required to gain sufficient quantities of material, can mean that the costs of plant-derived materials can be significantly higher than those of fossil-derived competitors. Targeted breeding and developments in biotechnology offer the potential to lower the costs associated with use of plant-derived materials.

The development of genomic sequencing, enabling linking of genes and their associated expressed products and traits, offers a significant new tool to enable further and wider exploitation of plants for industrial uses. This information can be exploited through either genetic modification (GM) or through non-GM routes by highlighting traits and targets for conventional breeding methods. It is anticipated that application of genomics to processing microorganisms as well as crops providing feedstocks, will significantly add to the value of developments in the plant sector (Smallwood, 2006).

Conventional breeding has developed oilseed rape (*Brassica napus*) cultivars that can accumulate long-chain fatty acids such as C20:1 and C22:1, however the ability to accumulate short-chain fatty acids is limited. Similarly the ability to accumulate industrially useful hydroxy fatty acids and epoxy fatty acids is also limited with conventional breeding methods. Due to its close relationship to the crucifer *Arabidopsis* and its associated characterised genome, and the relative ease with which genes can be inserted into *Brassica* species, oilseed rape is seen as a key target species for genetic manipulation.

Greater understanding of the gene and enzyme control of the fatty-acid synthesis pathway has led to the development of a number of modified rape cultivars, including high lauric (containing up to 30% lauric acid). Other ongoing targets include:

- High erucic (up to 90% erucic acid)
- High ricinoleic acid
- High petroselinic acid
- Very long-chain polyunsaturated fatty acids (C20, C22) (for replacement of fish oils in the nutritional supplement sector).

Plant oil metabolic pathways have also been studied to identify enzymes that may be of industrial interest in modifying or manipulating plant oils. Study of plants with unusual or novel oils has enabled identification of enzymes capable of catalysing the introduction of functional epoxy groups, acetylenic bonds and other useful modifications into fatty acids (Smallwood, 2006). These modifications significantly improve the industrial functionality of such feedstocks for use in applications such as polymer, paint, lubricant and ink formulations, reducing the need for primary industrial conversion steps such as epoxidation of fatty acids in the polymer sector.

However, the complexity of plant biochemical systems and the lack of knowledge relating to their interactions means that significant further work is still required in most cases to commercialise such developments (Smallwood, 2006). Similar problems affect the development of plants with enhanced levels of desirable secondary metabolites, where knowledge of routes of synthesis, substrates and rate-limiting processes is limited. In addition, complex biochemical interactions may be involved in their formation.

To try and enhance yields of speciality chemicals, attention has focused on use of plant trichomes, the glands directly producing or accumulating oils and chemicals that are typically used to protect plants from predation. Genetic and enzymic mapping of such cells has started to identify opportunities to enhance their productivity. Examples include enhancement of alkaloid biosynthesis in opium poppy trichomes (Ounaroon *et al.*, 2003).

Genomic technologies can also be used to direct development of new and novel products in crops. Development is currently underway to produce *Brassica* cultivars capable of producing the biodegradable polymers polyhydroxybutyrate (PHB) and polyhydroxybutyrate/hydroxyvalerate (PHB/V) directly in plants, rather than through microbial fermentation routes. Polyhydroxyalkanoates have also been expressed in transgenic *Arabidopsis* plants. Currently, microbial fermentation routes to produce these materials require significant energy input and production in plants could reduce process energy demands, adding to the environmental credentials of the resulting end products.

Genomic technologies also offer the potential to produce high-value proteins and peptides in plants for industrial and pharmaceutical uses, though extraction of such materials and ensuring they retain activity can be difficult. Work continues to develop transgenic bacteria as well as transgenic plants capable of synthesising new polymer forms. In either case, plants will have an important role in acting as feedstocks in any successful development.

Efforts in genomics have also been directed at manipulation of starch synthesis to modify the amylose:amylopectin ratio to increase susceptibility to α-amylase digestion (Smallwood, 2006), in this case to improve ethanol production via fermentation. Such developments would also have spin-offs for other industrial uses of starch where there is a preference for amylose or amylopectin forms.

In other areas, targets for manipulation include over-expression of oil-producing cells or fruiting in attempts to increase overall production of tissues that are metabolically active in producing the chemicals of interest, to enhance yields of useful chemical materials and reduce costs of production.

2.8.1   Protection of Conventional Food Crop Chains

Despite the potential advantages that GM technologies can offer, there has been a general reluctance to accept wide-scale use of genetically modified crops in some parts of the world, particularly in Europe, although GM maize (resistant to corn borer) is increasing in area in Europe. One of the concerns is cited as the potential risk of transfer of traits to neighbouring food crops or related species via gene transfer.

The development of crops with modified fatty-acid profiles, developed through conventional breeding programmes, for example high erucic acid crops, also pose a risk to food crop chains. As such there is a requirement to ensure there is little or no risk of cross-fertilisation and intermixing with conventional cultivars, and to ensure identity preservation and separation of crop produce. Such systems have been running effectively for decades.

Dealing with such problems and adopting methodologies to reduce any risk of crop contamination places additional labour, infrastructure and financial burdens on growers. To address such concerns it has been proposed that non-food crop plants unrelated to current food crops and native flora (to avoid risk of cross-pollination) should be used as potential hosts for engineered industrial use traits. Crambe, (*Crambe abyssinica*) has been identified as a suitable model oil crop plant (EPOBIO 2007). Crambe is a plant that has already been commercialised on a relatively small scale to exploit its high erucic acid content. Elsewhere, safflower has been proposed as a potential candidate, as well as the use of algae, moss and the aquatic plant duck weed in contained 'bioreactor' systems.

The ability to utilise such biotechnological solutions on a wider scale, to address some of the highlighted problems affecting use of plant-derived materials, will rely on confidence building within the general public where there is scepticism, and development of measures by plant breeders, agronomists and farmers to reduce any associated risks. However, despite these problems, restricted access to GM technologies will not restrict the ever-widening use of crop-derived feedstocks by industry, as there are other areas developing to help increase the opportunities for wider exploitation of plants as chemical feedstocks

2.8.2   Cell and Tissue Culture

Some plant cells can be encouraged to grow *in vitro* under specific controlled conditions, which can be used to produce secondary plant metabolites at higher yields than those found in plants, but at a high cost. There are few examples where the technological difficulties and costs involved have led to commercialisation currently, but the valuable pharmaceutical alkaloid berberine, the quinnone shikonin and the terpenoid paclitaxel are all potential candidates for production by cell culture.

Clearly such approaches are only suited to high-value outlets, but there are still other developing opportunities to access lower-value bulk-market applications for plant-derived materials.

2.8.3 Biorefining

The petrochemical industry typically works on a 'build-up' approach where the base oil feedstock is fractionated, and complex materials are built up from simpler ones, producing a wide array of materials in the process, for a range of market outlets. The future exploitation of plant materials is seen in a somewhat similar fashion, although in contrast to the petrochemical industry, there will typically be an initial breaking up of more complex materials into simpler building blocks that can then be utilised and built on with the support of chemical, biochemical and catalytic processes, to produce more complex products synonymous with those produced by today's petrochemical industry. This whole crop approach to industrial use of plant-derived material is typically termed biorefining (see Chapter 1).

Exploitation of plant-based products will increasingly entail removal of valuable products from plant-based sources before breaking up, or utilising, structural or storage metabolites as feedstocks for other processes. For example, cereal brans separated from grains destined for ethanol production can be used as sources of oils and phenolic compounds, and then exploited as an animal feed additive or alternatively can be used a fermentation stock for further ethanol production, while the grain endosperm is processed for its starch content and residual protein value to produce ethanol (or other fermentation products) and the residual distillers grain is separated into high-value protein and low-value animal feed fractions, or used as a fuel in the plant. The development of large-scale biofuel processing plants is helping to drive the commercial reality of such approaches and the first commercial steps in this direction are being made. Such developments are not new and reflect the processing industry attitude of finding a value for any by-product stream. What is changing is the recognition of the potential value of the components of such by-product streams and the new scale of operation that is making extraction of such materials more commercially viable

The development of complex enzyme systems to release cellulose and hemicellulose from lignin, and release their component sugars for ethanol fermentation, will lead to the development of biorefineries capable of utilising both traditional arable crops (for food or industrial uses) and their lignocellulosic waste products (such as straw or corn stover, brans and corn cobs). Such integrated ethanol fermentation plants are capable of producing a range of materials, including sugar syrups, lactic acid, citric acid, amino acids and starches or flours for food or non-food uses. Such integrated approaches will help to reduce current concerns related to pressures on land to produce for both our food and renewable raw material needs by optimising the use of plant-derived biomass and utilising materials currently regarded as waste in many cases.

2.8.4 Thermochemical Routes of Exploitation

In addition to biotechnology routes of exploitation, rapid thermochemical routes offer an alternative route to exploitation of biomass in its widest sense, being less affected by the material composition of the feedstock. Controlled fast pyrolysis of biomass leads to charring and the volatilisation of organic decomposition products and other compounds that can be condensed into so-called 'bio-oils', thick acidic oils containing a complex mixture of oxygenated compounds. While these can be used as fuels in their own right in static engines, and some useful materials can be derived from them including resins, they require further upgrading (through hydro-cracking) and conventional refining to provide a wide range of chemical building blocks similar to those found in fossil diesel. Alternatively biomass directly, or a bio-oil intermediary can be gasified to produce syngas (a mixture of hydrogen an carbon monoxide) that can be converted to fuels or catalysed to produce dis-tillates, including ethanol, propanol, butanol and other alcohols and many other commodity and speciality chemicals, providing almost limitless opportunities for chemical synthesis. The cost of gasification and bio-oil upgrading is a barrier to current wide-scale development, and currently large-scale processing is required to ensure viability, which has a significant impact on biomass demand. Further de-velopment and refinement in both pyrolysis and gasification is required to optimise the respective technologies and downstream use of products.

2.9 Concluding Comments

The examples given in this review show how key plant metabolites traditionally used to provide renewable raw materials for both the chemical and pharmaceutical sector predominantly fall into two camps:

1. Industries based on extraction of bulk quantities of raw material, which, with relatively little additional chemical modification, are put to industrial use at relatively low cost.
2. Industrial processes and supply chains where plants are grown specifically for the production of specific small volumes of high-value materials, for example for use in pharmacy, that are otherwise difficult to synthesise synthetically.

Despite the wide range of opportunities that have been highlighted, such ap-proaches severely limit the opportunities to which plant-derived materials can currently contribute. The first results in bulk production at lowest possible cost, which offers little, if any, additional reward to producers and results in wastage of plant biomass unless all by-product streams can be utilised. The second typi-cally results in small-scale opportunities for a small number of growers and often exposes growers to risk of over-production and price slumps that reduce grower confidence. Neither route encourages any significant investment in breeding and development.

Production for extraction of speciality chemicals directly from plants is likely to remain a niche operation, perhaps with biotechnology offering options to optimise individual plant yields and reduce land demands. However, development of both scientific understanding of plant metabolism and the technology available to exploit this, combined with environmental and fuel-security drivers facilitating production of large-scale fermentation and biorefining capabilities, offers a new opportunity to reconsider the role that plant-derived materials can offer the chemical industry. Crops that efficiently produce starch, sugars or oils are still key to the development of such industries and risk clashing with land demands for food crops unless opportunities to use non-food components are realised, and/or crops suited to more marginal or drought-prone areas are developed to reduce pressures on productive arable land areas. In the case of thermal routes of conversion, the origin of the plant biomass is immaterial and agricultural materials regarded as waste could become valuable chemical feedstocks providing returns to rural economies while reducing the impacts on land requirements.

There is still considerable work to be done to develop biorefining and thermochemical routes of plant exploitation. Isolating specific compounds in pure forms for industrial use from heterogeneous biomass presents a challenge analogous to that which faces the petrochemical industry. There will be a need for considerable cooperation and colocation of facilities to keep costs to a minimum, which means that new industrial partnerships are required to deliver on a large scale, to bring together wide-ranging interests in plant production, plant biotechnology, processing, extraction and separation engineering, enzymology and catalysis, chemistry, and thermo and chemical engineering, to help deliver a bio-based solution to the problems of depleting oil reserves and the need to reduce greenhouse gas emissions without impacting on what will be a growing demand for food.

References

Biavati, B., R. Piccaglia and M. Marotti, Biological Activity of Essential Oils, *Agroindustria* 2, 95–97 (2003).

BREW, *Medium and Long-Term Opportunities and Risks of the Biotechnological Production of Bulk Chemicals From Renewable Resources – The Potential of White Biotechnology, the BREW Project.* Final report, 2006.

Enhance, *Green Chemicals and Biopolymers From Rapeseed Meal With Enhanced End-Use Performance*, Final report of EU-funded Enhance project QLK5-1999-01442 (2003).

EPOBIO, *Oil Crop Platforms for Industrial Uses*, CPL Press, Newbury, UK (2007).

Fowler, M.W., Plants, Medicines and Man, *J. Sci. Food Agric.* 86, 1797–1804 (2006).

Guinda, A., M.C. Dobarganes, M.V. Ruiz-Mendez and M. Mancha, Chemical and Physical Properties of a Sunflower Oil with High Levels of Oleic and Palmitic Acids, *Eur. J. Lipid Sci. Technol.* 105, 130–137 (2003).

Hooper, L. and A. Cassidy, A Review of the Health Care Potential of Bioactive Compounds, *J. Sci. Food Agric.* 86, 1805–1813 (2006).

Ma, J.K.-C., Production of Recombinant Biopharmaceuticals in Plants – A Potential Solution for Global Health. In: *Proceedings XVI International Plant Protection Conference*, Vol. 2, 15–18 October 2007, Glasgow, Scotland, BCPC, Alton, UK, 528–529 (2007).

Morris, D., The Next Economy: From Dead Carbon to Living Carbon, *J. Sci. Food Agric.* 86, 1743–1446 (2006).

Ounaroon, A., G. Decker, J. Schmidt, F. Lottspeich and T.M. Kutchan, (R,S)-Reticuline 7-O-methyltransferase and (R,S)-Norcoclaurine 6-O-Methyltransferase of *Papaver somniferum*: cDNA Cloning and Characterisation of Methyl Transfer Enzymes of Alkaloid Biosynthesis in Opium Poppy, *Plant J.* **36**, 808–819 (2003).

Palmieri, S., Renewable Industrial Bio-Based Products from Agricultural Resources, *Agroindustria* **2**, 49–52 (2003).

Padley, F.B., F.D. Gunstone and J.L. Harwood, Occurrence and Characteristics of Oils and Fats, in *The Lipid Handbook* 2nd Edn, Gunstone, F.D., Harwood, J.L. and Padley, F.B., (Eds) 47–223, Chapman & Hall, London (1994).

Smallwood, M., The mpacts of Genomics on Crops for Industry. *J. Sci. Food Agri.* **86**, 1747–1754 (2006).

Turley, D.B., F.J. Areal and J.E. Copeland, *The Opportunities for Use of Esters of Rapeseed Oil as Bio-Renewable Solvents*, HGCA Research Review **52**, Home-Grown Cereals Authority, London (2004).

Turley, D.B., M.Froment and S. Cook (Eds) *Development of* Euphorbia Lagascae *as a New Industrial Oil Crop*, ADAS, Wolverhampton (2000).

Vink, E., *The Applications of Natureworks PLA*. Paper presented at Industrial Applications of Bioplastics Congress, 3–5 February, Central Science Laboratory, York, UK (2002).

3

Green Chemical Technologies

Francesca M. Kerton

Department of Chemistry, Memorial University of Newfoundland, Canada

3.1 Introduction

In the pursuit of chemicals from renewable resources, methods encompassing the principles of green chemistry and chemical engineering should be considered (Figures 3.1 and 3.2) (Anastas and Warner, 1998; Anastas and Zimmerman, 2003).

When using renewable feedstocks, one will always be adhering to some of these principles, as one of the twelve principles of green chemistry is that 'a raw material as feedstock should be renewable rather than depleting wherever technically and economically practicable;' and one of the twelve principles of green engineering is that 'material and energy inputs should be renewable rather than depleting'. However, it has been noted that these particular principles are currently little realized in the chemical industry, since production of organic chemicals from renewable feedstocks is often neither technically nor economically feasible (Lichtenthaler, 2002). However, with the increasing costs of ever-depleting fossil-based raw materials, the changeover of the chemical industry to renewable feedstocks is looming on a fast-approaching horizon and new clean methods for their transformation and processing must be developed. Many chemical and pharmaceutical companies now have teams of researchers focused on the implementation of clean technologies and the diversification of their feedstocks in order to become less dependent on oil and other non-renewable resources.

Introduction to Chemicals from Biomass Edited by James Clark and Fabien Deswarte
© 2008 John Wiley & Sons, Ltd

1. Prevention
It is better to prevent waste than to treat or clean up waste after it has been created.

2. Atom Economy
Synthetic methods should be designed to maximize the incorporation of all materials used in the process into the final product.

3. Less Hazardous Chemical Syntheses
Wherever practicable, synthetic methods should be designed to use and generate substances that possess little or no toxicity to human health and the environment.

4. Designing Safer Chemicals
Chemical products should be designed to effect their desired function while minimizing their toxicity.

5. Safer Solvents and Auxiliaries
The use of auxiliary substances (e.g., solvents, separation agents, etc.) should be made unnecessary wherever possible and innocuous when used.

6. Design for Energy Efficiency
Energy requirements of chemical processes should be recognized for their environmental and economic impacts and should be minimized. If possible, synthetic methods should be conducted at ambient temperature and pressure.

7. Use of Renewable Feedstocks
A raw material or feedstock should be renewable rather than depleting whenever technically and economically practicable.

8. Reduce Derivatives
Unnecessary derivatization (use of blocking groups, protection/deprotection, temporary modification of physical/chemical processes) should be minimized or avoided if possible, because such steps require additional reagents and can generate waste.

9. Catalysis
Catalytic reagents (as selective as possible) are superior to stoichiometric reagents.

10. Design for Degradation
Chemical products should be designed so that at the end of their function they break down into innocuous degradation products and do not persist in the environment.

11. Real-time analysis for Pollution Prevention
Analytical methodologies need to be further developed to allow for real-time, in-process monitoring and control prior to the formation of hazardous substances.

12. Inherently Safer Chemistry for Accident Prevention
Substances and the form of a substance used in a chemical process should be chosen to minimize the potential for chemical accidents, including releases, explosions, and fires.

Figure 3.1   The twelve principles of green chemistry

3.2   What are Green Chemistry and Green Engineering?

About fifteen years ago, the United State's Environmental Protection Agency (EPA) defined green chemistry as 'innovative chemical technologies that reduce or eliminate the use or generation of hazardous substances in the design, manufacture and use of chemical products'. Using the principles shown in Figures 3.1 and 3.2 as guidelines, chemists are able to evaluate and improve current procedures and develop new ones that will have a limited impact on the environment and therefore be more sustainable and economical in the long term. Many processes

1. Inherent Rather Than Circumstantial
Designers need to strive to ensure that all materials and energy inputs and outputs are as inherently nonhazardous as possible.

2. Prevention Instead of Treatment
It is better to prevent waste than to treat or clean up waste after it is formed.

3. Design for Separation
Separation and purification operations should be designed to minimize energy consumption and materials use.

4. Maximize Efficiency
Products, processes, and systems should be designed to maximize mass, energy, space, and time efficiency.

5. Output-Pulled Versus Input-Pushed
Products, processes, and systems should be "output pulled" rather than "input pushed" through the use of energy and materials.

6. Conserve Complexity
Embedded entropy and complexity must be viewed as an investment when making design choices on recycle, reuse, or beneficial disposition.

7. Durability Rather Than Immortality
Targeted durability, not immortality, should be a design goal.

8. Meet Need, Minimize Excess
Design for unnecessary capacity or capability (e.g., "one size fits all") solutions should be considered a design flaw.

9. Minimize Material Diversity
Material diversity in multicomponent products should be minimized to promote disassembly and value retention.

10. Integrate Material and Energy Flows
Design of products, processes, and systems must include integration and interconnectivity with available energy and materials flows.

11. Design for Commercial 'Afterlife'
Products, processes, and systems should be designed for performance in a commercial "afterlife."

12. Renewable Rather Than Depleting
Material and energy inputs should be renewable rather than depleting.

Figure 3.2   The twelve principles of green engineering

indeed involve several of the principles simultaneously. For example, in an attempt to develop a green process to synthesise natural fragrance compounds (see Figure 3.3), we recently combined catalytic chemistry and the use of an alternate solvent (i.e. supercritical carbon dioxide) to maximise atom economy (principles 2, 5 and 9) (Olsen *et al.*, 2006).

Several other principles were also encompassed, including principles 3 and 7, as we replace industrially used vinyl acetate (derived from petroleum, explosive, carcinogenic and neurotoxic) by acetic acid – a renewable and safer acyl donor.

Also, in the context of this book, it is worth noting that many products derived from renewable resources will be inherently biodegradable and therefore, adhere to principle 10.

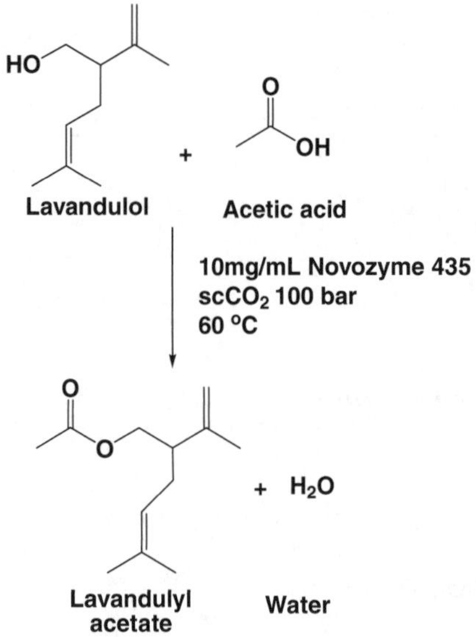

Figure 3.3   *Biocatalytic transformation of the natural terpenol, Lavandulol, in supercritical carbon dioxide (scCO$_2$)*

Researchers interested in renewable resources should use the principles listed in Figures 3.1 and 3.2 when designing new processes, products and materials. The principles focus your thinking in terms of sustainable design criteria and have already provided innovative solutions to a wide range of problems, and will ultimately benefit the environment, economy and society (otherwise known as the triple bottom line).

Green chemical technologies are important in the field of renewables in two main ways:

(1) The processing of the feedstock e.g. extraction using green solvents
(2) The transformations of the feedstocks e.g. catalytic alkene metathesis reactions.

Just as several principles of green chemistry are encompassed in a single process, sometimes these two areas overlap and a green chemical technology is simultaneously used in the processing and transformation of a renewable material e.g. sub- and supercritical water oxidation of biopolymers.

An overview of the main green chemical technologies that can be applied to renewable resources follows. Case studies from the literature will be presented focusing on solvent replacement, alternative energy sources and catalysis. This chapter will not cover all aspects of green chemistry and engineering – it is merely an introduction to this exciting, fast-moving area.

3.3 Evaluating the Environmental Effects of Chemistry and Green Metrics

To evaluate the 'greenness' of a chemical process, researchers must consider each step within it (Anastas and Warner, 1998).

For some of the steps outlined in Figure 3.4, the same essential question can be asked: Is this toxic to humans, animals and the environment? It would be near impossible to design a process that was 100% non-toxic and, therefore, a balance must be achieved between what is desired, what is technically possible and what is economic.

As well as addressing toxicity, green chemistry and green engineering seek to address the risks and hazards of chemical manufacture. This is possible through applying the concept of inherently safer design (ISD). At its core, ISD aims to avoid relying on mechanical safety devices and procedures. At the moment, chemical plants are designed to be safe, but most are not designed to be inherently safe (Lancaster, 2002). Many infamous chemical disasters, such as the incident at Bhopal, could have potentially been prevented with ISD. At the heart of that accident was the storage of methylisocyanate (MIC). Elsewhere, an alternative, safer route to the desired product, which did not require the storage of MIC, was already being used and therefore, the accident could have been avoided.

Often there is more that one alternative for a particular process and, therefore, metrics have been developed in order to compare different methods. The two most commonly used metrics are atom economy (AE) and E-Factor (see Figure 3.5). The most basic of these is atom economy, which is defined as the molecular weight of the desired product/(s) divided by the molecular weight of all of the reactants, and was developed by Barry Trost (Anastas and Warner, 1998; Lancaster, 2002). This metric is intended for use in the planning stages of process development, as no experimental data is necessary. While it does not take into account energy use, auxiliaries, catalysts or the toxicity of the waste, it does provide a simple approach to understanding where waste is generated. A similar metric – efficiency factor or E-factor – was developed by Roger Sheldon as the kg of waste produced per kg product. E-factor includes auxiliaries, solvents and excess reagents used. This metric is used during the experimental stage to compare the waste generated between two different routes. It often becomes apparent at this stage that the

1. Starting materials/feedstocks

2. Reaction types

3. Reagents

4. Solvents and reaction conditions

5. Chemical products/target molecules

Figure 3.4 *The five parts of a typical chemical process*

$$AE = \frac{MW \text{ Desired Product(s)}}{MW \text{ Reactants}} \times 100\%$$

$$E\text{-Factor} = \frac{Kg \text{ waste}}{Kg \text{ Product}}$$

Figure 3.5   *Green metrics*

single largest contributor to waste in a process is the solvent employed at each stage. Consequently, solvent replacement has become one of the most extensively investigated areas of green chemistry.

An award-winning example of green chemistry being used in the redesign of a process is in the production of Pfizer's antidepressant sertraline (see Figure 3.6) (Clark, 2002; Ritter, 2002; US Environmental Protection Agency).

In 2001, sertraline was the most prescribed drug of its class. The key improvement in the sertraline synthesis was reducing a three-step sequence in the original process to a single step. Overall, the new commercial process offers improved safety and material handling, reduced energy use, reduced waste (by 690 metric tons), reduced solvent requirement (from 60 000 gal to 6000 gal per ton of sertraline), reduced raw material usage, reduced water use and significantly increased overall yield. Remarkable improvements were, in particular, achieved by performing one-pot reactions, avoiding the isolation of a number of intermediates and through the introduction of catalysts with better chemoselectivity. In addition, the use of environmentally benign ethanol as the only solvent eliminated the need to use, distil and recover the various other solvents employed.

3.4   Alternative Solvents

Solvents are widely used in the chemical industry and play a variety of roles e.g. mass and heat transfer (Adams *et al.*, 2004). They are conventionally volatile organic compounds (VOC) that lead to significant emissions to the atmosphere and possess substantial risks such as flammability and toxicity. Some of these solvents are already banned in the pharmaceutical industry (e.g. benzene) and others

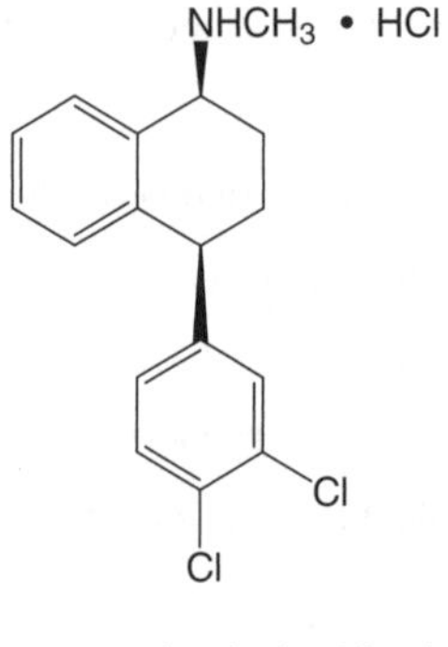

Figure 3.6   *Sertraline hydrochloride (Zoloft)*

should only be used if unavoidable (e.g. toluene and hexane) (US Department of Health and Human Services, 1997). Ironically, hexane, which is a hazardous air pollutant, is used to extract a wide range of vegetable oils. According to the EPA *Toxic Release Inventory*, more than 20 million kg of hexane are released into the atmosphere per year through these processes in the US alone (DeSimone, 2002). In contrast, green solvents are those that have low toxicity, can be easily recycled, are inert and do not contaminate the product. This includes water, heptane, ethyl acetate, ethanol and *tert*-butyl methyl ether. Other alternative (green) solvents currently being investigated by the green chemistry community include supercritical fluids, water, low-melting ionic liquids, fluorous solvents and liquid polymers e.g. poly(ethylene glycol). However, by far the best solvent is no solvent (or in reality, no auxiliary/added solvent), which can be an incredibly appealing option, especially when a reagent and/or product is a liquid.

3.4.1 Supercritical Fluids

In 2004, the academic Green Chemistry Presidential Award was given to Eckert and Liotta for their research on 'benign tunable solvents coupling reaction and separation processes' (US Environmental Protection Agency). They demonstrated that, in particular, supercritical fluids (SCFs) and expanded organic solvents can be used to perform a variety of novel organic reactions. In addition, the unique properties associated with these solvent systems mean that they can easily be used in the major two steps of any chemical processes, namely the reaction and the separation/purification (Eckert *et al.*, 2004). This reduces cost and energy usage, as well as waste production.

By definition, an SCF is a gas compressed to a pressure greater than its critical pressure (P_c) and heated to a temperature higher than its critical temperature (T_c). For example, the critical point for carbon dioxide occurs at a pressure of 73.8 bar and a temperature of 31.1 °C, as depicted in Figure 3.7. In this phase, regardless of the pressure applied, the fluid will not transcend to the liquid phase.

All SCFs offer the following advantages when used as a solvent (Jessop *et al.*, 1996; Kendall *et al.*, 1999):

(1) Misciblity with gases in all proportions above their T_c. This means that hydrogenations and other reactions involving gaseous reagents are enhanced in their selectivity and energy requirements.
(2) Weakening of solvation around the reacting species.
(3) Reduction of cage effects in radical reactions.
(4) Solvent is easily removed, due to its 'zero' surface tension, leaving the product in an easily processable, clean and solvent-free form.
(5) Recyclability and therefore near-zero waste production.

Many of these benefits result from SCFs having physical properties intermediate to those of gases and liquids. This is a simplistic (but useful) viewpoint as the properties of a substance may vary dramatically within the SCF phase boundaries,

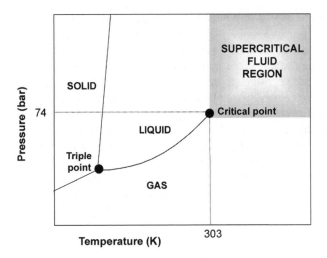

Figure 3.7   *Phase diagram for pure carbon dioxide*

as its temperature and pressure are changed (Clifford, 1998). Modification of the density, and hence solvent power, of this reaction medium with temperature and pressure may indeed afford some control over reaction pathways, resulting, in some cases, in product selectivity far superior to anything possible using conventional solvents.

There are two principal reactor types that have been used for reactions in SCF, as seen in Figure 3.8. Batch reactors can be readily equipped with a suitable window to assess homogeneity of the reaction mixture and are widely used in academic research.

In contrast, continuous flow reactors are already being used for hydrogenation reactions industrially (Licence *et al.*, 2003). They are simple to construct and modify, and possess excellent mass- and heat-transfer properties. In academia, flow reactors have been used in conjunction with a variety of heterogeneous catalysts to carry out many reactions, including hydrogenations, dehydrogenations, hydroformylations, Friedel–Crafts acylations and alkylations, etherifications and oxidations (Hyde *et al.*, 2001).

Supercritical Carbon Dioxide (scCO₂)

In many cases, CO_2 is seen as the most viable supercritical solvent. It is inexpensive and readily available (by-product of fermentation and combustion), non-toxic and non-flammable. It cannot be oxidised and therefore oxidation reactions using air or oxygen as the oxidant have been intensively investigated. In addition, it is inert to free-radical chemistry, in contrast to many conventional solvents.

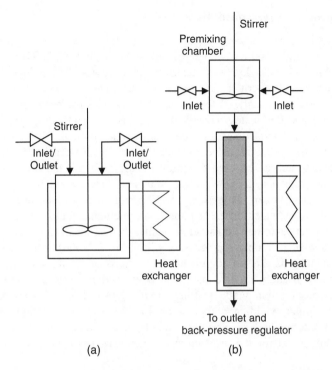

Figure 3.8   (a) Batch and (b) continuous-flow reactor for use with scCO₂

This has led to much research into free-radical-initiated polymerisations (Kendall *et al.*, 1999).

One of the most well known uses of scCO$_2$ is in the extraction of caffeine from coffee. This is carried out industrially on a huge scale producing many thousands of tons of decaffeinated coffee per year. New uses of scCO$_2$ for synthesis on an industrial scale are slowly becoming established, particularly where additional benefits, such as a higher purity product or a desirable physical property result (Beckman, 2004; Jessop and Leitner, 1999; Oakes *et al.*, 2001).

In addition to this, related areas such as liquid CO$_2$ and CO$_2$-expanded solvents should not be overlooked. Many additives and complex modifiers are being used to facilitate reactions in scCO$_2$ and perhaps the use of a small amount of organic solvent (perhaps from a bio-feedstock) could be justified in order to reduce the cost of a process and therefore lead to its uptake by industry. In addition to this, continued research into biphasic systems CO$_2$–water, CO$_2$–ionic liquids, CO$_2$–PEG/surfactants and CO$_2$–solids (including heterogeneous catalysts) is needed to deliver pure products and reduced cost to future end-users of this technology.

Near-critical and Supercritical Water

A wide, and increasing, range of synthetic reactions have been performed in near-critical (275 °C, 60 bar) and supercritical water (400 °C, 200 bar) (Savage, 1999; Siskin and Katritzky, 2001; Katritzky *et al.*, 2001). The unique dissociation properties of near-critical/supercritical water have been widely used to perform acid- and base-catalysed reactions, which negates the need for any added acid or base and eliminates subsequent neutralisation and salt disposal. More recently, it was demonstrated that microwave reactors could be used to perform even faster reactions in near-critical water (Leadbeater, 2005). Sub- and supercritical water can be used to simultaneously process and transform a renewable material in a green way, e.g. sub- and supercritical water oxidation (SCWO) of biopolymers. A view cell was recently used to look at the decomposition of wood under different conditions, including temperature, pressure and oxygen concentration, in order to gain a better understanding of SCWO and its potential in biomass processing (Shoji *et al.*, 2006). Also, using subcritical water – this time in conjunction with enzymatic catalysis – a new, environmentally friendly method has been developed for analysing the antioxidant content of onion waste (Turner *et al.*, 2006). This technology, which is quicker, higher yielding and uses up to 100 times less organic solvents than traditional methods, is another example of a technique that could potentially be employed for the production of chemicals from biomass on a large scale.

3.4.2   Water

In recent years, there has been considerable interest in chemistry in aqueous media and 'on water' (Li, 2005). Table 3.1 summarises all the advantages and disadvantages of this green medium.

Amazing advances have been made, whereby reactions that are typically considered unsuited to the presence of moisture, e.g. Grignard-type chemistry, can be performed in water. Also, many reactions show enhanced rates compared with related reactions in VOC solvents. Although studies are still in progress, it has been proposed that some of these reactions are not actually taking place in water

Table 3.1   *Advantage and disadvantages of using water as a solvent*

Advantages	Disadvantages
Opportunity for replacing VOCs	Distillation/Separation is energy intensive
Non-toxic, non-flammable, odourless, colourless, naturally occurring, inexpensive	Contaminated waste streams may be difficult to treat
High specific heat capacity (exothermic reactions can be more safely controlled)	High specific heat capacity (difficult to heat or cool rapidly)
Potential for easy catalyst recycling	

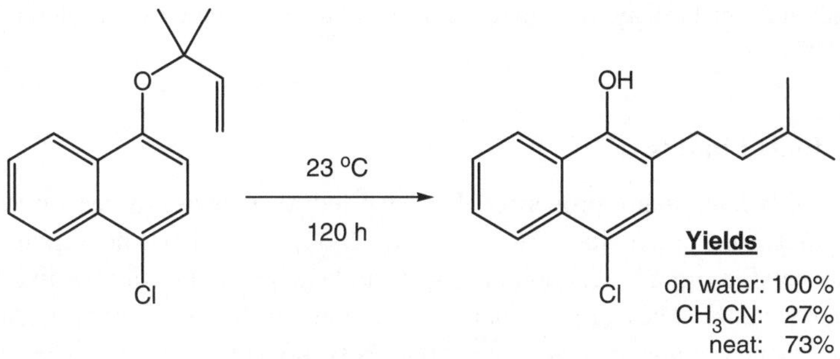

Figure 3.9   *Claisen rearrangement performed 'on water', in acetonitrile and neat*

but 'on water' as many organic substrates are hydrophobic and insoluble in this medium (Narayan *et al.*, 2005). An example of this phenomenon is shown in Figure 3.9.

Water as a reaction solvent can be also used as an effective method to separate homogeneous catalysts from a reaction mixture and allow them to be recycled and reused, and thus give higher turnover numbers and reduce waste. Several strategies have been developed for this purpose including the design of ligands that can make complexes preferentially dissolve in the aqueous phase (e.g. addition of $-SO_3^-Na^+$ groups), the addition of a phase transfer catalyst or suitable surfactant, and vigorous stirring. For example, kinetic resolution of secondary alcohols can be achieved using chiral Mn oxidation catalysts that are insoluble in water, in combination with a phase transfer catalyst (PTC) such as tetraethylammonium bromide (Sun *et al.*, 2003). In the absence of the PTC, the enantioselectivity is negligible, whereas an enantioselectivity of up to 88% can be achieved when the PTC was used. Another example that illustrates the remarkable potential of water as a solvent is the use of a water-soluble diruthenium complex as a recyclable catalyst in the aerobic oxidation of a range of primary and secondary alcohols (Komiya *et al.*, 2006). In this particular example, the catalyst-containing aqueous phase can be re-used three times without any loss in activity. In addition, as the rate of reaction is 14 times greater for the oxidation of primary alcohols compared with secondary alcohols, this strategy may provide a way of selectively converting primary alcohols to aldehydes in the presence of secondary and tertiary alcohols.

Many Lewis acid catalysts, such as aluminium chloride, are highly moisture sensitive and cannot be recycled and re-used. Therefore, on an industrial scale they can lead to considerable waste. However, one way to overcome this is by using a water-stable molecular Lewis acid such as ytterbium or scandium triflate (Ding *et al.*, 2006). It was recently discovered that rate enhancements could be achieved by adding small amounts of ligand to ytterbium triflate-catalyzed Michael addition reactions in water (Ding *et al.*, 2006). Although in this study a recycling study on the catalyst was not performed, it is likely that these species can be easily recycled in a similar way to related nitration catalysts previously reported (Waller *et al.*, 1997).

3.4.3   Ionic Liquids

Ionic liquids have many properties that have led to their use as reaction media and in materials processing (Adams *et al.*, 2004). They have no vapour pressure, so volatile organic reaction products can be separated easily by distillation or under vacuum. They are thermally stable and can be used over a wide temperature range compared with conventional solvents, and their properties can be readily adjusted by varying the anion and cation. For example, 1-butyl-4-methyl-imidazolium (bmim) tetrafluoroborate (BF_4) is a hydrophilic solvent, whereas its hexafluorophosphate (PF_6) analogue is hydrophobic. It has also been clearly demonstrated that the choice of ionic liquid can dramatically affect the outcome of a chemical reaction (Earle *et al.*, 2004). The reaction of toluene and nitric acid was performed in three different ionic liquids. Conversions and selectivities were excellent in each case, but the products were different i.e. oxidation occurred in one case and nitration in another. In general, ionic liquids can dissolve many metal catalysts without expensive modifications, as both species are ionic, or they themselves can act as the catalytic species. A wide range of catalytic reactions has been performed in these designer solvents, including: hydrogenations, C—C bond-forming reactions and biotransformations (Dupont *et al.*, 2002; Welton, 2004). In most cases, the ionic liquid-containing catalyst phase can be easily recycled.

As well as the most widely used ionic liquids, such as those containing the anions and cations shown in Figure 3.10, new, potentially more benign, ionic liquids are being developed based on non-toxic, degradable ions (Carter *et al.*, 2004). Ionic

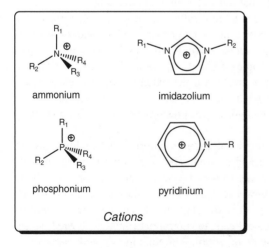

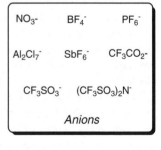

Figure 3.10   *Cations and anions commonly found in low-melting ionic liquids*

liquids containing anions derived from the sweeteners saccharin and acesulfame have properties similar to those containing the bis(trifluoromethyl)sulfonyl imide anion.

Acetylation reactions which could be of use in the transformation of bio-sourced alcohols and sugars have been performed in bmim-derived ionic liquids (Forsyth *et al.*, 2002; Alleti *et al.*, 2005). If the dicyanamide anion is incorporated into the liquid, mild acetylations of carbohydrates can be performed at room temperature, in good yields, without any added catalyst (Forsyth *et al.*, 2002). The ionic liquid acts as an efficient base catalyst.

Biopolymer processing and modification has been performed in ionic liquids, and in 2005, a US EPA Presidential Green Chemistry Challenge Award was given for the use of ionic liquids to dissolve and process cellulose for advanced new materials (US Environmental Protection Agency). It was found that cellulose from virtually any source (fibrous, amorphous, pulp, cotton, bacterial, filter paper etc.) can be dissolved readily and rapidly, without derivatisation, in a low-melting ionic liquid (IL), 1-butyl-3-methylimidazolium chloride ([C$_4$mim]Cl) by gentle heating (especially with microwaves) (Swatloski *et al.*, 2002). The dissolved polymer can be precipitated from water in controlled architectures (fibers, membranes, beads, flocs etc.) by a range of techniques. Blended and composite materials can also be formed by incorporating functional additives (Zhu *et al.*, 2006b). The additives can be soluble in the IL, e.g. dyes, or dispersed/insoluble e.g. nanoparticles. Notably, the IL can be recycled by at least two energy-saving methods. Other biopolymers can also be dissolved in [Bmim]Cl and ionic liquid solutions of chitin and chitosan can reversibly adsorb carbon dioxide (Xie *et al.*, 2006). The use of ionic liquids to dissolve naturally sourced material and the chemistry that can then be performed is wide-ranging and likely to expand dramatically in the coming years.

Related to ionic liquids are substances known as deep eutectic solvents or mixtures. A series of these materials based on choline chloride (HOCH$_2$CH$_2$NMe$_3$Cl) and either zinc chloride or urea have been reported (Abbott *et al.*, 2002; 2003). The urea/choline chloride material has many of the advantages of more well-known ionic liquids (e.g. low volatility), but can be sourced from renewable feedstocks, is non-toxic and is readily biodegradable. However, it is not an inert solvent and this has been exploited in the functionalisation of the surface of cellulose fibres in cotton wool (Abbott *et al.*, 2006). Undoubtedly, this could be extended to other cellulose-based materials, biopolymers, synthetic polymers and possibly even small molecules.

3.4.4 Other Alternatives to VOCs: 'Solventless', Biphasic and Bio-Sourced Solvents

Throughout the chemical industry, many products, such as poly(propylene) are made without the use of a solvent or are performed in the gas phase. In many cases, one of the reactants also acts as the solvent. As the field of green chemistry has grown, so has the number of so-called 'solventless' reactions, including

$$PTSA, MW$$

$$CH_3(CH_2)_3OH + CH_3(CH_2)_{14}COOH \xrightarrow{\text{160 °C, 5 min}} CH_3(CH_2)_{14}COO(CH_2)_3CH_3 + H_2O$$

$$93\%$$

Figure 3.11   *Solvent-free, acid catalysed esterification [PTSA = p-toluenesulfonic acid]*

transesterifications, condensations and rearrangements. These reactions are often performed in combination with other green chemistry tools, including microwave heating and photochemistry.

Microwave heating and catalysis have been successfully used in the solvent-free synthesis of cosmetic fatty esters (Villa *et al.*, 2003). Two kinds of reaction were performed; acid-catalysed esterification (Figure 3.11) and phase-transfer catalysed alkylations, both reactions affording near quantitative yields when microwave heating was used. It should be noted that diethyl ether and water were used in the purification of the product, and alternative purification/separation procedures would be required if this process was performed on an industrial scale, due to the flammability risk of diethyl ether.

Microwave heating has also been applied in the solvent-free phosphorylation of microcrystalline cellulose (Gospodinova *et al.*, 2002). In the isolation step of this procedure, only water and ethanol were used as additional solvents. Wax esters have been produced from vegetable oils using a solvent-free enzymatic process (Petersson *et al.*, 2005); this is particularly noteworthy as enzymes are often intolerant to high concentrations of substrates. The examples of solvent-free procedures described here show that solvents are not always required in the transformation of naturally sourced biopolymers and also in the chemistry of small molecules that can be obtained from a biorefinery.

Fluorous solvents, including perfluorocarbons, are chemically inert, thermally stable, non-flammable and, in many cases, non-toxic. In green chemistry, the field of fluorous biphasic catalysis has developed and is a useful way of taking advantage of the high selectivity and reactivity of homogeneous catalysts, whilst at the same time enabling their facile recycling (Horvath, 1998). The concept is illustrated in Figure 3.12. Most ligands frequently found in homogeneous catalysts have been modified with perfluorinated ponytails, e.g. $P[CH_2CH_2(CF_2)_5CF_3]_3$, to effectively partition the catalyst in the fluorous phase when the reaction is complete. At this time, fluorous biphasic chemistry is not performed on an industrial scale, due to the expense of the solvents and designer catalysts. Thermoregulated biphasic catalysts that do not require a special solvent have also been developed (Bergbreiter, 2002). The ligands are modified with low to medium molecular weight polymer chains, the phase in which the catalyst resides (i.e. its solubility in a particular solvent) is dependent on the temperature of the system, and the catalyst will precipitate from the reaction mixture at either low or high temperatures. These systems combine many of the benefits of homogeneous catalysis with the ease of separation typically associated with heterogeneous catalysts.

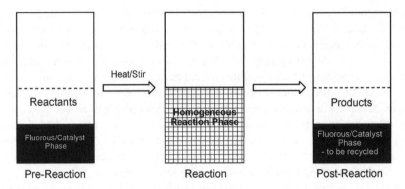

Figure 3.12   *Fluorous biphasic systems*

A wide range of solvents can be potentially sourced from renewable feedstocks, including alcohols such as ethanol and esters such as ethyl lactate. These should be relatively easily incorporated into processes as they have the properties traditionally associated with commonly used laboratory solvents e.g. volatility. In the preparation of magnetic tape coatings, ethyl lactate, which is non-toxic, biodegradable and not an air pollutant, has been used to replace butan-2-one (MEK), 4-methylpentan-2-one (MIBK) and toluene (Nikles *et al.*, 2001). These solvents pose a toxicity risk to workers, they are hazardous air pollutants and are derived from petroleum. There is clearly a wide range of situations that could also use such relatively benign solvents, including the processing of renewable feedstocks.

3.5   Energy Considerations: Microwaves, Ultrasound, Electricity and Light

In addition to a solvent, most reactions require additional energy to occur quickly and successfully. Energy requirements can be reduced by using a catalyst, but most reactions are also heated to allow them to occur over a short enough time frame. As energy is becoming ever more expensive, chemists are looking at other ways to provide the energy required in chemical processes, including microwaves, ultrasound, electricity and light. It is worth noting that whereas microwaves and ultrasound typically accelerate conventional, thermal reactions, some reactions can only be performed under electrochemical or photochemical conditions and therefore these techniques truly expand the chemistry toolbox beyond that which is normally considered.

3.5.1   Microwave-Assisted Chemistry

The use of microwave heating in order to effect chemical transformations has grown rapidly in the last 10 years and is being widely used in combinatorial/ high-throughput approaches (Kappe, 2004; Hoz *et al.*, 2005; Nüchter *et al.*, 2004;

Roberts and Strauss, 2005). As was described earlier, an added advantage to microwave chemistry is that often no solvent is required. In recent years, many commercial reactors have come on the market and some are amenable for scaling up reactions to the 10 kg scale. These new instruments allow direct control of reaction conditions, including temperature, pressure, stirring rate and microwave power, and therefore, more reproducible results can be obtained. For most successful microwave-assisted reactions, a polar solvent that is able to absorb the energy and efficiently convert it to heat is required, however, even solvents such as dioxane that are more or less microwave transparent can be used if a substrate, coreagent or catalyst absorbs microwaves well. In fact, ionic liquids have been exploited in this field as polar additives for low-absorbing reaction mixtures.

Reactions performed using microwave heating include metal-catalysed C—C and C—N bond formations, condensations, cycloadditions, rearrangements, oxidations, glycosylations, substitutions and radical chemistry. In the field of polymers derived from renewables, microwaves have been used to assist in the ring-opening polymerisation of an oxazoline monomer obtained from soy (Hoogenboom and Schubert, 2006). A well-defined polymer via a living polymerisation has been obtained both in acetonitrile and in the absence of any additional solvent. This bulk polymerisation achieved completion in only eight minutes. Importantly, the fatty-acid side chains of the polymer did not react and would therefore be available for further chemistry if required. A recent publication shows great promise for the use of microwave heating in the processing and transformation of biopolymers. Wood liquefaction of low-value materials was achieved rapidly and quantitatively under microwave irradiation in the presence of glycols or anhydrides and phosphoric acid (5–20%) as a catalyst (Krzan and Kunaver (2006). The efficiency of liquefaction increased with high microwave power, longer radiation time and higher acid concentration. However, even in the absence of the acid catalyst, 50% liquefaction could be achieved. Microwave heating has also been used to increase the rate of conversion of lignin (a renewable source of aromatic molecules) into vanillin, a product with commercial value (Clark *et al.*, 2006). Although yields of product were not increased dramatically, this study clearly shows the dramatic rate enhancement that microwaves can provide, as maximum conversion was achieved after 15 min of microwave heating compared to 24 h of conventional heating.

3.5.2 Sonochemistry

Ultrasonic energy (use of a sonicator) is finding use in chemical transformations, often improving rates and selectivity. Ultrasound refers to sound with a frequency higher than those detectable by the human ear, typical frequencies used in chemical applications are 20–100 kHz. In general, most commercial ultrasound laboratory equipment is suitable for performing chemical reactions, although they are sold as cleaning baths or as cell disruptors (immersion probes/horns) (Kardos and Luche, 2001). The energy applied and the temperature of a reaction is more easily controlled using an immersion horn emitter, more details on experimental procedures and instruments can be found in the chemical literature. The chemical effects of ultrasound result from the phenomenon of cavitation (Cintas and Luche, 1999). Gas-filled microbubbles grow and eventually collapse; this generates very high local temperatures ($\sim$5000 °C) and pressures (over 1000 bar). Flashes of light known as sonoluminescence may also accompany the collapse of the bubbles. There are several theories as to the origins of the heat, pressure and light generated during cavitation, and an explanation of these is beyond the scope of this chapter. However, it is clear that interesting chemistry can be performed under these unusual conditions.

A wide range of reactions has been performed using sonication, including substitutions, additions, reductions, organometallic chemistry and materials synthesis (Cintas and Luche, 1999). In heterogeneous systems, the mechanical effects of cavitation enhance the rates of reaction in a similar fashion to rapid agitation of the reaction mixture. Also, when metal surfaces are involved, the ultrasound can have a cleaning effect and reduce surface passivation. In homogeneous reactions, electron transfer and radical formation are favoured compared to conventional methods, and therefore reactions that typically involve these show interesting behaviour when performed using ultrasound e.g. oxidations. As shown in Figure 3.13, sonication accelerates the epoxidation of long-chain fatty esters with m-chloroperbenzoic acid (MCPBA) (Cintas and Luche, 1999). Reactions using sugars (and carbohydrates) as reagents, including glycosylation with long-chain alcohols, can be performed (Kardos and Luche, 2001). Therefore, it is likely that sonochemistry will

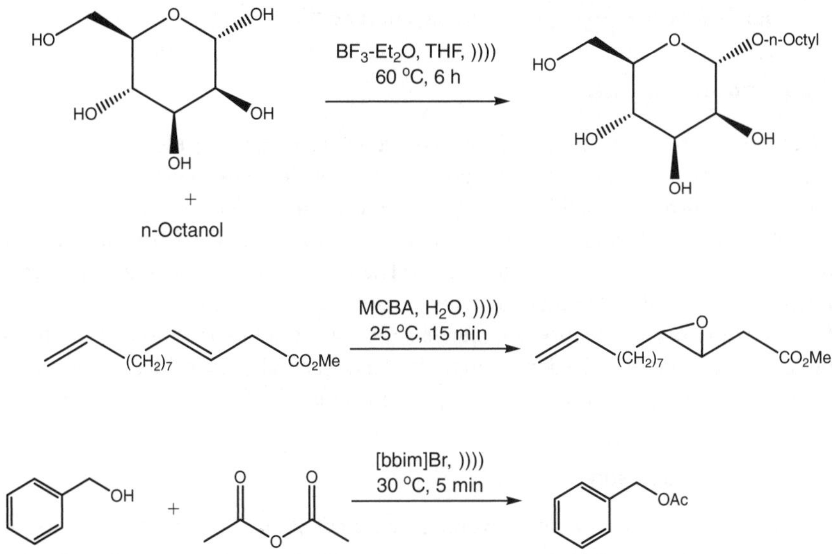

Figure 3.13   Selected sonochemical reactions)))) = ultrasound]

continue to play an important role in the development of the diverse chemistry of carbohydrates and therefore, renewable resources. Ultrasound can also be used in conjunction with other tools from the green chemistry toolbox, such as alternative solvents. The conversion of a wide range of alcohols to esters using acetic anhydride as the acetylating agent have been performed under ultrasonic irradiation in the absence of any added catalysts using a room temperature ionic liquid as the medium and also the promoter for the reaction (Gholap *et al.*, 2003). The esters could be easily separated from the reaction mixture because of the involatile nature of the ionic solvent. Some of the alcohols studied included sugars, terpenols and fatty alcohols that could potentially be derived from biomass. It was proposed that the ionic liquid, as well as providing the medium for the reaction, was also acting as an acid catalyst.

Sonication has been widely used in biochemical and microbiological laboratories to aid in the digestion of cells and other naturally sourced materials. Therefore, it is not surprising that sonochemistry is useful in the extraction of both primary and secondary metabolites from renewable feedstocks. For example, it has been used in the extraction of chitin from prawn shells (Kjartannson *et al.*, 2006). The properties of the chitin extracted using this method were different to those produced using conventional methods. The chitin extracted under sonication had a lower protein content and decreased crystallinity. The use of ultrasound has been compared with conventional methods in the extraction of carvone and limonene from caraway seeds (Chemat *et al.*, 2004). Extraction rates were up to twice as fast using ultrasound and additionally, the yield of carvone was higher and it was suggested that this would be useful in reducing separation costs. It was also noted that the temperature of extraction was important and that lower temperature, ultrasound-assisted extractions afforded a purer extract.

3.5.3 Electrochemistry

Electrochemistry is a well-established technique in industry, e.g. electroplating and chlorine production, and could be used to give novel bio-sourced products as, often, unusual chemistry is possible. Other green credentials of electrochemistry include use of aqueous solvents, energy efficiency and atom efficiency (replacement of reagents with electrons) (Lancaster, 2002). There is currently considerable research interest in fuel cells and other electrochemistry-based renewable energy sources. However, the use of electrochemical reactors in the transformation of platform molecules and other bio-sourced materials has generally been investigated to a lesser extent and may provide exciting opportunities for future investigation.

3.5.4 Photochemistry

Photochemical processes have several green credentials. Photons are very clean reagents, leaving no residues. Many often use fewer raw materials. Since energy is more directed, reaction temperatures are generally low. This may give higher selectivities, by reducing by-product formation from competing reactions. Also, some reactions proceed by reaction pathways that are only possible photochemicallly and not thermally. However, there are several problems that have limited their uptake by industry. Light sources are expensive and fragile, so equipment costs are higher compared to thermal processes. Much of the energy from a lamp is wasted in the form of heat or just by being the wrong wavelength for a particular process. Therefore, overall energy efficiency is reduced in many systems. Reactor fouling occurs where the photochemical window or wall becomes covered in a compound and prevents efficient transfer of the photons to the bulk of the reaction mixture.

Photochemistry is used commercially in the synthesis of Vitamin A and Vitamin D_3, Figure 3.14. Vitamin A is formed via a photochemical isomerisation process and the synthesis of Vitamin D_3 involves a photochemical electrocyclic ring-opening step. The industrial relevance of this process is due to the lack of a viable thermal alternative and therefore, if unusual and useful chemistry is only possible via photochemistry, an industrial process may result, due to the high intrinsic value of the chemical product.

In an interesting twist on photochemistry, sunlight has been used to perform acylation and oxygenation reactions on the kg scale with near quantitative yields (Oelgemöller *et al.*, 2005). The substrates are commercially available starting materials and products are key intermediates for industrial applications.

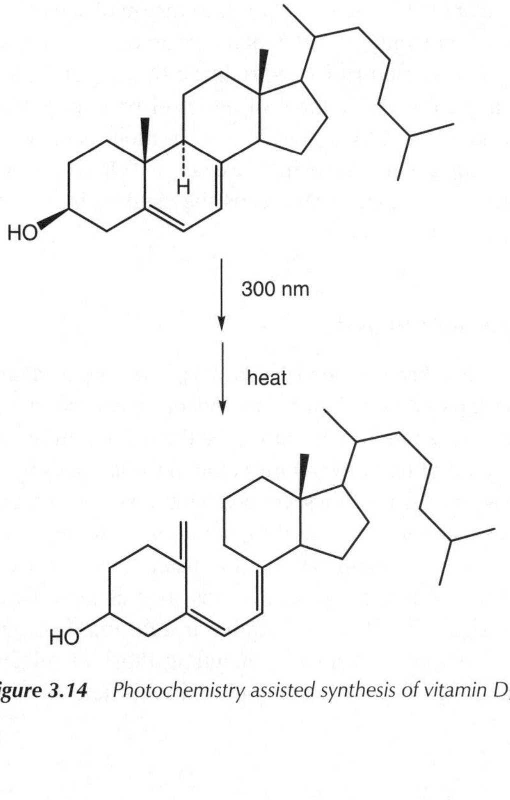

Figure 3.14 *Photochemistry assisted synthesis of vitamin D₃*

3.6   Catalysts

Approximately 90% of all commercial chemicals come into contact with a catalyst at some point in their manufacture. A catalyst is commonly defined as 'a material which changes the rate of attainment of chemical equilibrium without itself being changed or consumed in the process' (Lancaster, 2002). The activation energy of a reaction is lowered and therefore, the overall energy required for a process is significantly reduced. In addition to this, selectivity in reactions is generally increased. Therefore, it is not surprising that 'catalytic chemistry rather than stoichiometric chemistry' is one of the mantras of green chemistry. There is no area where this is more apparent than oxidation catalysis, which could also play a major role in the transformation of renewable feedstocks. Historically, organic chemists use stoichiometric metal reagents as oxidising agents. These generate considerable waste and in many cases, e.g. chromium, this is highly toxic. To overcome this, many researchers are developing catalysts that can be used in very small amounts alongside more benign oxidants such as hydrogen peroxide or air. These catalysts, as with catalysts for all reactions, can be homogeneous (same phase as the substrates) or heterogeneous (different phase to the substrates).

In addition to these two broad classifications, there are other sub-classes that are important in green chemistry. These include asymmetric catalysts, biocatalysts, phase-transfer catalysts and photocatalysts. In assessing the effectiveness of a catalyst, three parameters are considered: selectivity, turnover frequency (TOF) and turnover number (TON). Selectivity is the amount of substrate converted to the desired product as a percentage of the total consumed substrate. A low selectivity means that a significant amount of a by-product is forming and that means increased waste production. TOF is the number of moles of product produced per mole of catalyst per second. If the TOF is low, longer reaction times or increased catalyst quantity will be required and these increase costs. TON is the amount of product per mole of catalyst. A large TON means the catalyst is robust and has a long lifetime, reducing waste and cost.

3.6.1   Homogeneous Catalysts

As shown in Table 3.2, homogeneous catalysts possess both advantages and disadvantages when compared with their heterogeneous counterparts. The short service life and difficulty in separation are probably the most significant disadvantages. Researchers are attempting to overcome some of these issues by developing more tolerant, stable species such as the second generation Grubbs catalysts. Significant efforts have been made to improve the separation of homogenous catalysts and products including development of dendritic species that can be used in membrane reactors and the use of novel solvent systems such as ionic liquids and fluorous materials (Gladyz, 2002). The considerable insight that homogeneous catalysts can provide on reaction mechanisms, including those of related heterogeneous systems, has led to significant research efforts in this field.

Table 3.2   *Comparison of heterogeneous and homogeneous catalysts*

Heterogeneous	Homogeneous
Usually distinct solid phase	Same phase as reaction medium
Facile separation	Often difficult to separate that can lead to product contamination
Readily regenerated and recycled	Expensive/difficult to recycle
Lower activity and selectivity	High rates and selectivity
May be diffusion limited	Not diffusion controlled
Quite sensitive to poisons	Usually robust to poisons
Long service life	Short service life
Often high-energy process	Often takes place under mild conditions
Continuous processing is possible	
Poor mechanistic understanding	Often mechanism well understood

Olefin or alkene metathesis of seed oils and their fatty-acid methyl esters has been an active area of research in renewables for some time, using either heterogeneous or homogeneous catalysis. However, due to the amazing functional group tolerance of recent homogeneous catalysts, fundamental studies can be performed and ways to make the process economically viable can be determined. The Dow Chemical Company recently reported on the mechanism of ethenolysis using methyl oleate as a model substrate (Figure 3.15); this involved extensive experimental kinetic studies and also computational investigations (Burdett *et al.*, 2004).

Catalytic alkene metathesis chemistry has recently been combined with an additional step in a single pot to further modify natural oils. A metathesis-isomerisation-methoxycarbonylation-transesterification reaction sequence has been performed to yield high-value oxygenates (Zhu *et al.*, 2006a). A palladium catalyst is added to the reaction mixture once maximum conversion in the metathesis step is achieved, and it is heated under 400 psi of carbon monoxide.

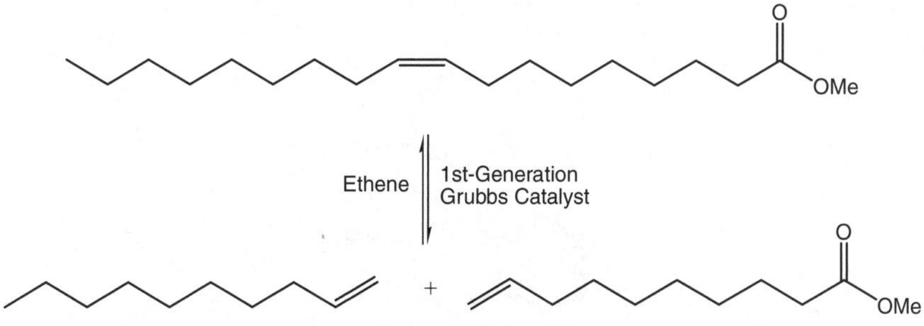

Figure 3.15   *Transformation of a seed oil based feedstock, methyl oleate, using catalytic olefin metathesis*

Other homogeneous catalytic systems that are of relevance to the field of renewables include catalysts for selective deoxygenation of sugar polyols (Schlaf, 2006), and for the production of polyesters and polycabonates via ring-opening polymerisation of lactides and through using carbon dioxide as a feedstock (Chisholm and Zhou, 2004). Catalysts have been developed over the last 100 years to functionalise hydrocarbon substrates e.g. hydroformylation of α-olefins, and therefore, new catalysts and a new understanding of fundamental reaction steps needs to be established for the multifunctionalised substrates, including seed oils, available from nature. Initial studies show acid catalysts and late transition-metal complexes (Ru, Rh, Ir, Pd or Pt) in combination can be used to transform polyols. For example, a homogeneous Ru catalyst selectively deoxygenates 1,2-propanediol to *n*-propanol in the presence of triflic acid and hydrogen (Schlaf *et al.*, 2001). Excellent selectivity and moderate conversion levels are achieved and such systems show great promise for the future.

Asymmetric catalysis allows chemicals to be manufactured in their enantiomerically pure form and reduces derivatisation and multiple purification steps that would otherwise be required. The 2001 Nobel Prize was awarded for two of the most important asymmetric reactions; hydrogenations and oxidations. A variety of ligands suitable for asymmetric reductions are available commercially including BINAP, Figure 3.16. A BINAP Rh complex is used in the commercial production of 1-menthol to enantioselectively hydrogenate an alkene bond (Lancaster, 2002). Ru BINAP complexes can be used in asymmetric reductions of carbonyl groups (Noyori, 2005; Noyori and Hashiguchi, 1997).

3.6.2   Heterogeneous Catalysts

Many catalysts used in industry are heterogeneous, e.g. zeolites in the cracking of heavy crude oil. The actual reaction takes place on the surface of the solid, with the substrates and products being in the gas or liquid phase, depending on the type of reaction and reactor being used. Many involve a metal on some kind of support e.g. Pd on charcoal, Ni on alumina. The surface area and porosity of heterogeneous catalysts are important in determining their efficiency. Some of the

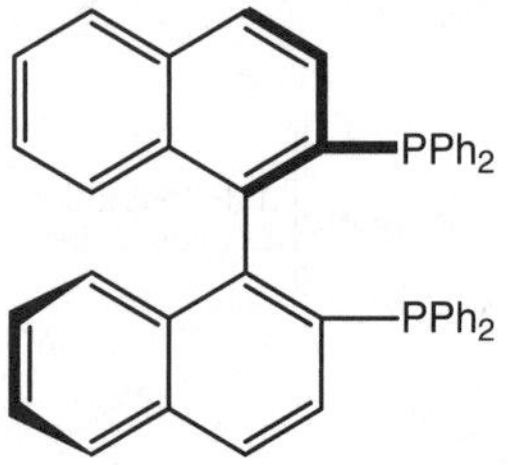

Figure 3.16   *The important chiral ligand, BINAP [2,2'-bis(diphenylphosphino)-1,1'-binaphthyl], S-enantiomer shown*

most successful heterogeneous catalysts are based on zeolites, as their pore size and distribution can be controlled in their preparation. Many simple organic reactions are base catalysed and heterogenised bases on polystyrene or silica supports can be used. Use of such catalysts reduces salt and/or organic waste streams. Large-pore (mesoporous) silicate materials have been particularly successful in this area (Clark et al., 2006).

Interestingly, biopolymers can be used in place of silica, alumina, other metal oxides and synthetic organic polymers as supports for heterogeneous catalysts. Chitosan has been investigated to the greatest extent due to the nitrogen functionalities of the glucosamine residues. A review of this field was recently published (Macquarrie and Hardy, 2005). The amine groups within the polymer can be reacted to give a range of new materials including chitosan–Schiff Bases. These sites can act as ligands for catalytic metal centres. Reactions studied include C—C bond formation, hydrogenation, oxidation and polymerisation. Chitosan has also been used to immobilise enzymes and may have applications in the food industry. Expanded starch has been chemically modified to give acidic, e.g. $-SO_3H$, or basic, e.g. NH_2, sites on its surface. These materials have been successfully employed in Knoevenagel and Michael reactions (Doi et al., 2002), and show potential as porous, organophilic heterogeneous catalysts. Expanded starches can be converted to mesoporous carbon by controlled pyrolysis; these new materials are called starbons (Budarin et al., 2006), and have a wide range of potential uses, including catalysis.

Heterogeneous catalysts have been successfully used in conjunction with alternative solvents, in particular $scCO_2$. For example, three types of heterogeneous catalyst were recently studied in the esterification of 2-ethylhexanoic acid and 2-ethyl-1-hexanol in $scCO_2$. Amberlyst® 15, a strong acid catalyst, did not show any activity. However, it did catalyse the dehydration of the alcohol to an alkene. A heterogeneous, Lewis-acid catalyst based on zirconia showed moderate activity and a supported enzyme, Novozym 325, showed limited activity (Ghaziaskar et al., 2006). A homogeneous Rh catalyst for hydroformylation was modified for immobilisation on silica and used for catalysis in $scCO_2$ (Meehan et al., 2000. The catalyst was found to be quite robust and its activity did not drop dramatically upon heterogenisation. The higher reaction rate was possibly due to the enhanced mass-transport properties of $scCO_2$ and its low viscosity. Studies on the homogeneous catalyst were performed in toluene and the solubility of the gaseous reagents would be significantly lower in this conventional solvent. Taking advantage of the high solubility of gases in $scCO_2$, hydrogenation of isophorone is being performed on an industrial scale in this alternative solvent, using a 2% Pd catalyst in a continuous-flow reactor within a multipurpose plant. A multipurpose plant is designed so the catalyst can be changed in order to perform a different type of reaction, e.g. alkylation or esterification, if required (Licence et al., 2003). Continuous-flow reactors using $scCO_2$ as the solvent have also been used for the kinetic resolution of alcohols using enzymes in an ionic liquid as the stationary, catalytic phase (Lozano et al., 2002; Reetz et al., 2003).

3.6.3   Biocatalysts

Biocatalysts (enzymes and whole cells or organisms) constitute an important area of green chemistry. Several recent US EPA Presidential Green Chemistry Challenge Awards have been given to researchers in this area: in 2006, for biocatalysis in the greener production of a cholesterol reducing drug; in 2005, for the development of enzymes for the interesterification of vegetable oils; and in 2004, for a new enzyme technology to improve paper recycling (US Environmental Proection Agency). So what are biocatalysts and enzymes? Biocatalysis is defined as the use of a biological system to catalyse the conversion of a single material to a defined product (Clark and Macquarrie, 2002). More specifically, certain proteins found within cells catalyse metabolic functions for the organism. These proteins, called enzymes, can be isolated, purified and used to catalyse organic reactions. While biocatalysis can be performed using whole cells, most green chemists use commercially available, isolated enzymes. There are six main classes of enzymes, their names indicating the type of reaction they catalyze (Table 3.3). Within each class there are sub-classes. For example, lipases is a subclass of hydrolases, which catalyze reactions involving the hydrolysis of water-insoluble esters (Bornscheuer and Lazlauskas, 1999). Biocatalysis offers a method of performing a chemical reaction under relatively mild conditions, including low temperatures and aqueous solvents, while providing high selectivity towards substrates and products. Enzymes are biodegradable, non-toxic, non-carcinogenic, and water soluble (Matlack, 2001). While enzymes are the perfect green catalyst in many regards, some inherent problems limit their widespread industrial use. They may be unstable under extreme conditions, the isolation of pure enzyme is difficult and expensive and reactions may have to be run in dilute solutions (Matlack, 2001). Considerable amounts of research have focused on combining supercritical fluids and enzymes, as both the catalyst and the solvent are environmentally benign and can provide high selectivities in reactions. Reactions studied include enantioselective esterifications, oxidations, asymmetric reductions, hydrolyses and carboxylations (Matsuda *et al.*, 2004). As enzymes are not soluble in SCFs, dispersion of the free enzyme potentially allows simple separation without the need for immobilisation. The structure and therefore function of the enzyme

Table 3.3   Classes of enzymes

Enzyme class	Catalyzed Reaction
Oxidoreductases	Oxidation –Reduction
Transferases	Group transfer
Hydrolases	Hydrolytic
Lyases	Reactions involving a double bond
Isomerases	Reactions involving isomerisation
Ligases	Join two molecules coupled with the breakdown of a phosphate bond

may be affected in a number of ways by $scCO_2$, which will influence its performance.

New reactor designs and immobilisation methods have been used to extend the lifetime of lipases in $scCO_2$ (Lozano *et al.*, 2004). Ceramic membranes have been coated with hydrophilic polymers and the enzyme covalently attached to these. In $scCO_2$, activities and selectivities were excellent and the half-life of the catalyst was enhanced. It is thought the hydrophilic layer of the membrane protected the enzyme. Operational stability of enzymes has also been increased by using ionic liquid/$scCO_2$ biphasic systems (Lozano *et al.*, 2002; Reetz *et al.*, 2003).

Enzymatic esterifications can also be performed under solventless conditions. Ethyl palmitate, found in chamomile and liquorice root, can be used to treat a variety of diseases (Kumar *et al.*, 2004). A higher yield is obtained for the biocatalytic synthesis of ethyl palmitate under solvent-free conditions compared to $scCO_2$, however, this is due to higher substrate and enzyme concentrations. In $scCO_2$, lower catalyst loading, higher reaction rates and easier downstream processing are possible. Therefore, if the process were scaled up, $scCO_2$ would be preferred to a solvent-free approach.

Lipases have also been used in hyphenated extraction-reaction routes to deriving natural oils using two high-pressure units in series. Canola oil was extracted from canola flakes and fatty acid ethyl esters synthesised using an immobilised enzyme (Lipozyme M) (Kondo *et al.*, 2002). It was clear that each unit in this process, whether reaction or extraction, needed to be run under its own optimum conditions.

An interesting method for recycling degradable polyesters using enzymes has been reported. Novozym 435, an immobilised lipase from *Candida antartica*, was used in $scCO_2$ to convert poly(ε-caprolactone) into repolymerisable oligomers (Matsumura *et al.*, 2001). The same enzyme can also be used in $scCO_2$ to reform the polymer. More enzymes will be developed in the future to degrade newly developed polymers from renewable feedstocks.

3.7   Conclusions

The field of chemistry can be thanked for the discovery of innumerable products that are responsible for the quality of life we enjoy today. Life-saving drugs, new energy technologies and highly durable polymers are only a few examples. In the past, chemists generally developed synthetic routes with one goal in mind: the product. Rarely did one consider the whole process to notice the irony that the synthesis of a cancer treatment drug may generate carcinogenic emissions. The persistence of chemicals in the biosphere and in our bodies is becoming a major global health issue. The vast majority of organic chemicals are made from depleting (non-renewable) feedstocks. Chemists and, importantly, chemical companies have begun actively searching for greener alternatives to their current manufacturing practices. Significant progress has been made in several key research areas, such as catalysis and environmentally benign solvents, and these should be used in the development of a new chemical industry based on renewable feedstocks. Outreach activities, both locally and globally, within the Green Chemistry community can highlight the potential for chemistry to solve many of the global environmental challenges we now face and its role in sustainable development (Poliakoff and Noda, 2004). The growth of green chemistry, especially in its application to renewable feedstocks, over the course of the past decade has been dramatic, but needs to be increased if we are to meet the challenges of sustainability.

References

Abbott, A.P., G. Capper, D.L. Davies, R.K. Rasheed and V. Tambyrajah, Quaternary Ammonium Zinc- or Tin-Containing Ionic Liquids: Water Insensitive, Recyclable Catalysts for Diels–Alder Reactions, *Green Chemistry*, **4**, 24–26 (2002).

Abbott, A.P., G. Capper, D.L. Davies, R.K. Rasheed and V. Tambyrajah, Novel Solvent Properties of Choline Chloride/Urea Mixtures, *Chemical Communications*, 70–71 (2003).

Abbott, A.P., T.J. Bell, S. Handa and B. Stoddart, Cationic Functionalisation of Cellulose Using a Choline Based Ionic Liquid Analogue, *Green Chemistry*, **8**, 784–786 (2006).

Adams, D.J., P.J. Dyson and S.J. Taverner, *Chemistry in Alternative Reaction Media*, John Wiley & Sons, Ltd, Chichester (2004).

Alleti, R., W.S. Oh, M. Perambuduru, Z. Afrasiabi, E. Sinn and V.P. Reddy, Gadolinium Triflate Immobilized in Imidazolium Based Ionic Liquids: A Recyclable Catalyst and Green Solvent for Acetylation of Alcohols and Amines, *Green Chemistry*, **7**, 203–206 (2005).

Anastas, P.Y. and J.C. Warner, *Green Chemistry: Theory and Practice*, Oxford University Press, New York (1998).

Anastas, P.Y. and J.B. Zimmerman, Design Through the 12 Principles of Green Engineering, *Environmental Science & Technology*, **37**, 94A–101A (2003).

Beckman, E.J., Using CO_2 to Produce Chemical Products Sustainably, *Environmental Science & Technology*, **36**, 347A–353A (2002).

Beckman, E.J., Supercritical and Near-Critical CO_2 in Green Chemical Synthesis and Processing, *Journal of Supercritical Fluids*, **28**, 121–191 (2004).

Bergbreiter, D.E., Using Soluble Polymers To Recover Catalysts and Ligands, *Chemical Reviews*, **102**, 3345–3384 (2002).

Bornscheuer, U.T. and R.J. Kazlauskas, *Hydrolases in Organic Synthesis*, Wiley-VCH, Weinheim, Germany (1999).

Budarin, V., J.H. Clark, J.J.E. Hardy, R. Luque, K. Milkowski, S.J. Tavener and A.J. Wilson, Starbons: New Starch-Derived Mesoporous Carbonaceous Naterials with Tunable Properties, *Angewandte Chemie International Edition*, **45**, 3782–3786 (2006).

Burdett, K.A., L.D. Harris, P. Margl, B.R. Maughon, T. Mokhtar-Zadeh, P.C. Saucier and E.P. Wasserman, Renewable Monomer Feedstocks Via Olefin Metathesis: Fundamental Mechanistic Studies of Methyl Oleate Ethenolysis with the First-Generation Grubbs Catalyst, *Organometallics*, **23**, 2027–2047 (2004).

Carter, E.B., S.L. Culver, P.A. Fox, R.D. Goode, I. Ntai, M.D. Tickell, R.K. Traylor, N.W. Hoffman and J.H. Davis Jr., Sweet Success: Ionic Liquids Derived from Non-Nutritive Sweeteners, *Chemical Communications*, 630–631 (2004).

Chemat, A., A. Lagha, H. AitAmar, P.V. Bartels and F. Chemcat, Comparison of Conventional and Ultrasound-Assisted Extraction of Carvone and Limonene from Caraway Seeds, *Flavour and Fragrance Journal*, **19**, 188–195 (2004).

Chisholm, M.H. and Z. Zhou, New Generation Polymers: The Role of Metal Alkoxides as Catalysts in the Production of Polyoxygenates, *Journal of Materials Chemistry*, **14**, 3081–3092 (2004).

Cintas, P. and J.L. Luche, Green Chemistry – The Sonochemical Approach, *Green Chemistry*, **1**, 115–125 (1999).

Clark, J.H., Editorial - Green All The Way Through!, *Green Chemistry*, **4**, G28 (2002).

Clark, J.H., V. Budarin, F.E.I. Deswarte, J.J.E. Hardy, F.M. Kerton, A.J. Hunt, R. Luque, D.J. Macquarrie, K. Milkowski, A. Rodriguez, O. Samuel, S.J. Tavener, R.J. White and A.J. Wilson, Green Chemistry and the Biorefinery: A Partnership for a Sustainable Future, *Green Chemistry*, **8**, 853–860 (2006).

Clark, J.H. and D.J Macquarrie (eds), *Handbook of Green Chemistry and Technology*, Blackwell Science, London (2002).

Clark, J.H., D.J. Macquarrie and S.J. Tavener, The Application of Modified Mesoporous Silicas in Liquid Phase Catalysis, *Dalton Transactions*, 4297–4309 (2006).

Clifford, A.A., *Fundamentals of Supercritical Fluids*, Oxford University Press, Oxford, UK (1998).

DeSimone, J.M., Practical Approaches to Green Solvents, *Science*, **297**, 799–803 (2002).

Ding, R., K. Katebzadeh, L. Roman, K.-E. Bergquist and U.M. Lindstrom, Expanding the Scope of Lewis Acid Catalysis in Water: Remarkable Ligand Acceleration of Aqueous Ytterbium Triflate Catalyzed Michael Addition Reactions, *Journal of Organic Chemistry*, **71**, 352–355 (2006).

Doi, S., J.H. Clark, D.J. Macquarrie and K. Milkowski, New Materials Based on Renewable Resources: Chemically Modified Expanded Corn Starches as Catalysts for Liquid Phase Organic Reactions, *Chemical Communications*, 2632–2633 (2002).

Dupont, J., R.F. de Souza and P.A.Z. Suarez, Ionic Liquid (Molten Salt) Phase Organometallic Catalysis, *Chemical Reviews*, **102**, 3667–3691 (2002).

Earle, M.J., S.P. Katdare and K.R. Seddon, Paradigm Confirmed: The First Use of Ionic Liquids to Dramatically Influence the Outcome of Chemical Reactions, *Organic Letters*, **6**, 707–710 (2004).

Eckert, C.A., C.L. Liotta, D. Bush, J.S. Brown and J.P. Hallett, Sustainable Reactions in Tunable Solvents, *Journal of Physical Chemistry B*, **108**, 18108–18118 (2004).

Forsyth, S.A., D.R. MacFarlane, R.J. Thomson and M. von Itzstein, Rapid, Clean, and Mild O-Acetylation of Alcohols and Carbohydrates in an Ionic Liquid, *Chemical Communications*, 714–715 (2002).

Ghaziaskar, H.S., A. Daneshfar and L. Calvo, Continuous Esterification or Dehydration in Supercritical Carbon Dioxide, *Green Chemistry*, **8**, 576–581 (2006).

Gholap, A.R., K. Venkatesan, T. Daniel, R.J. Lahoti and K.V. Srinivasan, Ultrasound Promoted Acetylation of Alcohols in Room Temperature Ionic Liquid Under Ambient Conditions, *Green Chemistry*, **5**, 693–696 (2003).

Gladyz, J.A. (Ed.), Thematic Issue on Recoverable Catalysts and Reagents, *Chemical Reviews*, **102**, 3215–3892 (2002).

Gospodinova, N., A. Grelard, M. Jeannin, G.C. Chitanu, A. Carpov, V. Thiery and T. Besson, Efficient Solvent-Free Microwave Phosphorylation of Microcrystalline Cellulose, *Green Chemistry*, **4**, 220–222 (2002).

Hoogenboom, R. and U.S. Schubert, Microwave-Assisted Cationic Ring-Opening Polymerization of a Soy-Based 2-Oxazoline Monomer, *Green Chemistry*, **8**, 895–899 (2006).

Horvath, I.T., Fluorous Biphase Chemistry, *Accounts of Chemical Research*, **31**, 641–650 (1998).

Howdle, S., *http://www.nottingham.ac.uk/~pczctg/Video_Clip_5.htm*. (2004), Department of Chemistry, University of Nottingham, UK.

Hoz, A.D.L., A. Díaz-Ortiz and A. Moreno, Microwaves in Organic Synthesis. Thermal and Non-Thermal Microwave Effects, *Chemical Society Reviews*, **34**, 164–178 (2005).

Hyde, J.R., P. Licence, D. Carter and M. Poliakoff, Continuous Catalytic Reactions in Supercritical Fluids, *Applied Catalysis A – General*, **222**, 119–131 (2001).

Jessop, P.G., Y. Hsiao, T. Ikariya and R. Noyori, Homogeneous Catalysis in Supercritical Fluids: Hydrogenation of Supercritical Carbon Dioxide to Formic Acid, Alkyl Formates, and Formamides, *Journal of the American Chemical Society*, **118**, 344–355 (1996).

Jessop, P.G. and W. Leitner, *Chemical Synthesis Using Supercritical Fluids*, Wiley-VCH, Weinheim, Germany (1999).

Kappe, C.O., Controlled Microwave Heating in Modern Organic Synthesis, *Angewante Chemie International Edition*, **43**, 6250–6284 (2004).

Kardos, N. and J.L. Luche, Sonochemistry of Carbohydrate Compounds, *Carbohydrate Research*, **332**, 115–131 (2001).

Katritzky, A.R., D.A. Nichols, M. Siskin, R. Murugan and M. Balasubramanian, Reactions in High-Temperature Aqueous Media, *Chemical Reviews*, **101**, 837–892 (2001).

Kendall, J.L., D.A. Canelas, J.L. Young and J.M. DeSimone, Polymerizations in Supercritical Carbon Dioxide, *Chemical Reviews*, **99**, 543–563 (1999).

Kjartannson, G.T., S. Zivanovic, K. Kristbergsson and J. Weiss, Sonication-Assisted Extraction of Chitin from Shells of Fresh Water Prawns, *Journal of Agriculture and Food Chemistry*, **54**, 3317–3323 (2006).

Komiya, N., T. Nakae, H. Sato and T. Naota, Water-Soluble Diruthenium Complexes Bearing Acetate and Carbonate Bridges: Highly Efficient Catalysts for Aerobic Oxidation of Alcohols in Water, *Chemical Communications*, 4829–4831 (2006).

Kondo, M., K. Rezaei, F. Temelli and M. Goto, On-Line Extraction-Reaction of Canola Oil with Ethanol by Immobilized Lipase in Supercritical Carbon Dioxide, *Industrial & Engineering Chemistry Research*, **41**, 5770–5774 (2002).

Kumar, R., G. Madras and J. Modak, Enzymatic Synthesis of Ethyl Palmitate in Supercritical Carbon Dioxide, *Industrial & Engineering Chemistry Research*, **43**, 1568–1573 (2004).

Krzan, A. and M. Kunaver, Microwave Heating in Wood Liquefaction *Journal of Applied Polymer Science*, **101**, 1051–1056 (2006).

Lancaster, M., *Green Chemistry: An Introductory Text*, Royal Society of Chemistry, Cambridge (2002).

Leadbeater, N.E., Fast, Easy, Clean Chemistry by Using Water as a Solvent and Microwave Heating, *Chemical Communications*, 2881–2902 (2005).

Li, C.-J., Organic Reactions in Aqueous Media with a Focus on Carbon–Carbon Bond Formations: A Decade Update, *Chemical Reviews*, **105**, 3095–3165 (2005).

Licence, P., J. Ke, M. Sokolova, S.K. Ross and M. Poliakoff, Chemical Reactions in Supercritical Carbon Dioxide: From Laboratory to Commercial Plant, *Green Chemistry*, **5**, 99–104 (2003).

Lichtenthaler, F.W., Unsaturated O- and N-Heterocycles From Carbohydrate Feedstocks, *Accounts of Chemical Research*, **35**, 728–737 (2002).

Lozano, P., T. de Diego, D. Carrie, M. Vaultier and J.L. Iborra, Continuous Green Biocatalytic Processes Using Ionic Liquids and Supercritical Carbon Dioxide, *Chemical Communications*, 692–693 (2002).

Lozano, P., G. Villora, D. Gomez, A.B. Gayo, J.A. Sanchez-Conesa, M. Rubio and J.L. Iborra, Membrane Reactor with Immobilized Candida Antarctica Lipase B for Ester Synthesis in Supercritical Carbon Dioxide, *Journal of Supercritical Fluids*, **29**, 121–128 (2004).

Macquarrie, D.J. and J.J.E. Hardy, Applications of Functionalized Chitosan in Catalysis, *Industrial & Engineering Chemistry Research*, **44**, 8499–8520 (2005).

Matlack, A.S., *Introduction to Green Chemistry*, Marcel Dekker, Inc., New York (2001).

Matsuda, T., T. Harada and K. Nakamura, Organic Synthesis Using Enzymes in Supercritical Carbon Dioxide, *Green Chemistry*, **6**, 440–444 (2004).

Matsumura, S., H. Ebata, R. Kondo and K. Toshima, Organic Solvent-Free Enzymatic Transformation of Poly(Epsilon-Caprolactone) into Repolymerizable Oligomers in Supercritical Carbon Dioxide, *Macromolecular Rapid Communications*, **22**, 1326–1329 (2001).

Meehan, N.J. A.J. Sandee, J.N.H. Reek, P.C.J. Kamer, P. van Leeuwen and M. Poliakoff, Continuous, Selective Hydroformylation in Supercritical Carbon Dioxide Using an Immobilised Homogeneous Catalyst, *Chemical Communications*, 1497–1498 (2000).

Narayan, S., H. Muldoon, M.G. Finn, V.V. Fokin, H.C. Kolb and K.B. Sharpless, 'On Water': Unique Reactivity of Organic Compounds in Aqueous Suspension, *Angewandte Chemie International Edition*, **44**, 3275–3279 (2005).

Nikles, S.M., M. Piao, A.M. Lane and D.E. Nikles, Ethyl Lactate: A Green Solvent for Magnetic Tape Coating, *Green Chemistry*, **3**, 109–113 (2001).

Noyori, R., Pursuing Practical Elegance in Chemical Synthesis, *Chemical Communications*, 1807–1811 (2005).

Noyori, R. and S. Hashiguchi, Asymmetric Transfer Hydrogenation Catalyzed by Chiral Ruthenium Complexes, *Accounts of Chemical Research*, **30**, 97–102 (1997).

Nüchter, M., B. Ondruschka, W. Bonrath and A. Gum, Microwave Assisted Synthesis – A Critical Technology Overview, *Green Chemistry*, **6**, 128–141 (2004).

Oag, R.M., P.J. King, C.J. Mellor, M.W. George, J. Ke and M. Poliakoff, Probing the Vapor–Liquid Phase Behaviors of Near-Critical and Supercritical Fluids Using a Shear Mode Piezoelectric Sensor, *Analytical Chemistry*, **75**, 479–485 (2003).

Oakes, R.S., A.A. Clifford and C.M. Rayner, The Use of Supercritical Fluids in Synthetic Organic Chemistry, *Journal of the Chemical Society – Perkin Transactions 1*, 917–941 (2001).

Oelgemöller, M., C. Jung, J. Ortner, J. Mattay and E. Zimmermann, Green Photochemistry: Solar Photooxygenations with Medium Concentrated Sunlight, *Green Chemistry*, **7**, 35–38 (2005).

Olsen, T., F. Kerton, R. Marriott and G. Grogan, Biocatalytic Esterification of Lavandulol in Supercritical Carbon Dioxide Using Acetic Acid as the Acyl Donor, *Enzyme and Microbial Technology*, **39**, 621–625 (2006).

Petersson, A.E.V., L.M. Gustafsson, M. Nordblad, P. Börjesson, B. Mattiasson and P. Adlercreutz, Wax Esters Produced by Solvent-Free Energy-Efficient Enzymatic Synthesis and Their Applicability as Wood Coatings, *Green Chemistry*, **7**, 837–843 (2005).

Poliakoff, M. and I. Noda, Plastic Bags, Sugar Cane and Advanced Vibrational Spectroscopy: Taking Green Chemistry to the Third World, *Green Chemistry*, **6**, G37–G38 (2004).

Reetz, M.T., W. Wiesenhofer, G. Francio and W. Leitner, Continuous Flow Enzymatic Kinetic Resolution and Enantiomer Separation Using Ionic Liquid/Supercritical Carbon Dioxide Media, *Advanced Synthesis & Catalysis*, **345**, 1221–1228 (2003).

Ritter, S.K., Green Challenge, *Chemical and Engineering News*, **80**, 26–30 (2002).

Roberts, B.A. and C.R. Strauss, Toward Rapid, 'Green', Predictable Microwave-Assisted Synthesis, *Accounts of Chemical Research*, **38**, 653–661 (2005).

Sarbu, T., T. Styranec and E.J. Beckman, Non-Fluorous Polymers with Very High Solubility in Supercritical CO_2 Down to Low Pressures, *Nature*, **405**, 165–168 (2000).

Savage, P.E., Organic Chemical Reactions in Supercritical Water, *Chemical Reviews*, **99**, 603–622 (1999).

Schiel, C., M. Oelgemöller, J. Ortner and J. Mattay, Green Photochemistry: The Solar-Chemical Photo–Friedel–Crafts Acylation Of Quinones, *Green Chemistry*, **3**, 224–228 (2001).

Schlaf, M., Selective Deoxygenation of Sugar Polyols to Alpha,Omega-Diols and Other Oxygen Content Reduced Materials – A New Challenge To Homogeneous Ionic Hydrogenation And Hydrogenolysis Catalysis, *Dalton Transactions*, 4645–4653 (2006).

Schlaf, M., P. Ghosh, P.J. Fagan, E. Hauptman and R.M. Bullock, Metal-Catalyzed Selective Deoxygenation of Diols to Alcohols, *Angewandte Chemie International Edition*, **40**, 3887–3890 (2001).

Siskin, M. and A.R. Katritzky, Reactivity of Organic Compounds in Superheated Water: General Background, *Chemical Reviews*, **101**, 825–836 (2001).

Shoji, D., N. Kuramochi, K. Yui, H. Uchida, K. Itatani and S. Koda, Visualized Kinetic Aspects of a Wood Block in Sub- and Supercritical Water Oxidation, *Industrial & Engineering Chemistry Research*, **45**, 5885–5890 (2006).

Sun, W., H. Wang, C. Xia, J. Li and P. Zhao, Chiral-Mn(Salen)-Complex-Catalyzed Kinetic Resolution of Secondary Alcohols in Water, *Angewandte Chemie International Edition*, **42**, 1042–1044 (2003).

Swatloski, R.P., S.K. Spear, J.D. Holbrey and R.D. Rogers, Dissolution of Cellulose with Ionic Liquids, *Journal of the American Chemical Society*, **124**, 4974–4975 (2002).

Turner, C., P. Turner, G. Jacobson, K. Almgren, M. Waldeback, P. Sjoberg, E.N. Karlsson and K.E. Markides, Subcritical Water Extraction and Beta-Glucosidase-Catalyzed Hydrolysis of Quercetin Glycosides in Onion Waste, *Green Chemistry*, **8**, 949–959 (2006).

US Department of Health and Human Services, Food and Drug Administration, and Center for Drug Evaluation and Research (CDER). *http://www.fda.gov/cder/guidance/Q3Cfinal.htm, Guidance for Industry, Q3C Impurities: Residual Solvents* (1997).

US Environmental Protection Agency. *http://www.epa.gov/greenchemistry/pubs/pgcc/past.html*.

Villa, C., E. Mariani, A. Loupy, C. Grippo, G.C. Grossi and A. Bargagna, Solvent-Free Reactions as Green Chemistry Procedures for the Synthesis of Cosmetic Fatty Esters, *Green Chemistry*, **5**, 623–626 (2003).

Waller, F.J., A.G.M. Barrett, D.C. Braddock and D. Ramprasad, Lanthanide(III) Triflates as Recyclable Catalysts for Atom Economic Aromatic Nitration, *Chemical Communications*, 613–614 (1997).

Welton, T., Ionic Liquids in Catalysis, *Coordination Chemistry Reviews*, **248**, 2459–2477 (2004).

Xie, H.B., S.B. Zhang and S.H. Li, Chitin and Chitosan Dissolved in Ionic Liquids as Reversible Sorbents of CO_2, *Green Chemistry*, **8**, 630–633 (2006).

Zhu, Y., J. Patel, S. Mujcinovic, W.R. Jackson and A.J. Robinson, Preparation of Terminal Oxygenates from Renewable Natural Oils by a One-Pot Metathesis-Isomerisation-Methoxycarbonylation-Transesterification Reaction Sequence, *Green Chemistry*, **8**, 746–749 (2006a).

Zhu, S.D., Y.X. Wu, Q.M. Chen, Z.N. Yu, C.W. Wang, S.W. Jin, Y.G. Ding and G. Wu, Dissolution of Cellulose with Ionic Liquids and Its Application: A Mini-Review, *Green Chemistry*, **8**, 325–327 (2006b).

4

Production of Chemicals from Biomass

Apostolis A. Koutinas, C. Du, R.H. Wang and Colin Webb

Satake Centre for Grain Process Engineering, School of Chemical Engineering and Analytical Science, University of Manchester, UK

4.1 Introduction

Oil and natural gas are currently the predominant raw materials used for the production of around 95% of worldwide chemical production. Figure 4.1 presents predominant production routes of various chemicals that are derived from the primary building blocks produced from crude oil (ethylene, propylene, butadiene, butenes, benzene, toluene, xylenes and methane). One of the major characteristics of these primary building blocks is that they are hydrocarbons, containing no oxygen or nitrogen in their molecular formula. These platform molecules could be used as end-products (e.g. benzene as solvent), monomers for polymer synthesis (e.g. ethylene, propylene) or as precursors for chemical production through addition of elements such as oxygen or nitrogen.

The imminent depletion of petroleum, increased public awareness and the effect of the emission of greenhouse gasses to global climate will eventually cause the substitution of petrochemical processes for biomass-based processes for chemical production. Koutinas *et al.* (2007a) introduced the term biochemurgy, that is the science that exploits fundamental principles from biochemical engineering, biochemistry and chemistry to develop clean and sustainable technologies that combine physical, chemical and biological processing to convert agricultural or

Introduction to Chemicals from Biomass Edited by James Clark and Fabien Deswarte
© 2008 John Wiley & Sons, Ltd

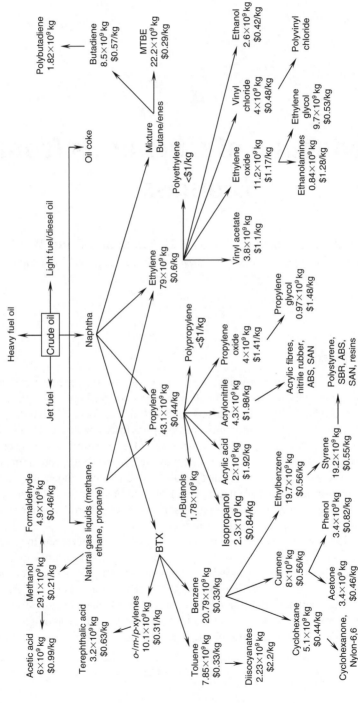

Figure 4.1 *Predominant production routes, capacities and unit cost of major organic chemicals and plastics (adapted from Webb et al., 2004)*

organic raw materials into chemical products. The materialisation of such technologies will not be easy to accomplish, but we should consider that biomass is the only raw material for sustainable production of chemicals. Energy could be derived from many different sources, but chemicals could be derived only from biomass.

The sustainable development of biomass-based processes would be dependent on the optimal utilisation or conversion of all biomass components (e.g. carbohydrates, protein, oil, lignin, functional minor constituents) into energy and chemical products. Treating all biomass components as the primary building blocks for future chemical production would create a completely new chemical industry, improve process economics, and minimise waste production. The substitution of hydrocarbons as the current petrochemically derived building blocks used for chemical production by oxygen- or nitrogen-containing molecules would be dependent on the production yield during microbial fermentation and/or chemical conversion, renewable resource availability and effect of renewable feedstock cost (Koutinas *et al.*, 2004a). In many cases, the development of renewable routes using biomass for chemical production may offer technological and environmental advantages in comparison to (petro)chemical processes.

In this chapter, major routes for the production of bulk chemicals from biomass components including carbohydrates, oils, proteins and minor constituents will be identified. These major production pathways rely on microbial fermentation, enzymatic transformation and green chemical extraction or synthesis.

4.2   Carbohydrates

Carbohydrates would be the predominant raw materials for future biorefineries. The major polysaccharides found in nature are cellulose, hemicellulose and starch (see Chapter 1). These molecules would be mainly utilised after they are broken down to their respective monomers via enzymatic hydrolysis, thermochemical degradation or a combination of these two. Cellulose and hemicellulose, together with lignin, constitute the main structural components of biomass. Starch is the major constituent of cereal crops. This section would focus on the potential utilisation of carbohydrates and lignocellulosic biomass for chemical production.

4.2.1   Chemical Production from Saccharides

The monossacharides glucose, xylose and arabinose, which could be predominantly derived from cellulose, starch and hemicellulose could be converted to various bulk chemicals through bioconversion, chemical modification or processing routes combining the previous two technologies. As crude oil and natural gas is to the petrochemical industry, glucose is the most important raw material to bioindustry. On an industrial scale, glucose is currently produced by the enzymatic hydrolysis of starch, which is derived from corn, wheat, potato and tapioca (Patel *et al.*, 2006). In 1998, the annual glucose production was over 15 million tonnes worldwide (Wilke 1999). Moreover, glucose could also be generated from the hydrolysis of cellulose contained in lignocellulosic biomass. Kim and Dale (2004) estimated that the annual crop residues and wasted crops from corn, barley, oat, rice, wheat, sorghum and sugar cane could be as much as 1500 and 73.9 million tonnes, respectively. If only 1% of the above biomass sources could be hydrolysed to saccharides, the total potential biochemical production would be over 3.5 million tonnes from only non-food raw materials, which is more than 10 times the current annual lactic acid production.

Fermentative conversion of saccharides into platform chemicals would be dependent on the development of viable biorefineries (Koutinas *et al.*, 2004a; Koutinas *et al.*, 2004b; Webb *et al.*, 2004; Koutinas *et al.*, 2006; Koutinas *et al.*, 2007a; Koutinas *et al.*, 2007b; Koutinas *et al.*, 2007c; Koutinas *et al.*, 2007d). A generic strategy leading to fermentative conversion of various biomass resources into platform chemicals is shown in Figure 4.2. In this process, the raw materials are processed to pure sugar solutions, sugar mixtures or nutrient-complete generic feedstocks. These fermentation media are then fermented by case-specific microorganisms to overproduce metabolic products. After purification, these metabolic products could be used as platform molecules for the production of various bulk chemicals through biological and/or chemical processing routes.

In one of the most comprehensive studies available in the literature, Werpy and Petersen tested 300 chemicals derived from renewable resources to identify the ones that could constitute the future biobased building blocks for chemical production (Werpy and Petersen, 2004). The 12 platform molecules that the report highlighted as the most promising ones are three 1,4-diacids (succinic, fumaric and malic), 2,5- furandicarboxylic acid, 3-hydroxy-propionic acid, aspartic acid, glucaric acid, glutamic acid, itaconic acid, levulinic acid, 3-hydro- xybutyrolactone, glycerol, sorbitol and xylitol/arabinitol. From these platform molecules, succinic acid, fumaric acid, malic acid, glutamic acid, itaconic acid and aspartic acid could be produced via microbial fermentations using glucose-based media, and natural or mutated microorgansims (Kim and Ruy, 1982; Yahiro *et al.*, 1997; Cao, 1997; Zeikus *et al.*, 1999; Taing and Taing, 2007). Genetically engineered microorganisms could be used for improved production of the above platform chemicals as well as the fermentative production of 3-hydroxypropionic acid (Tullo, 2007) employing predominantly glucose-based media (Werpy and Petersen, 2004). The platform molecules 2,5-furandicarboxylic acid, glucaric acid, levulinic acid, 3-hydroxybutyrolactone and sorbitol could be mainly produced by chemical conversion of biomass-derived saccharides (Werpy and Peterson, 2004). Glycerol would be available as a by-product of biodiesel production plants. Xylitol and arabinitol could be produced via fermentative conversion of xylose and arabinose, respectively.

Table 4.1 presents potential platform chemicals and their possible derivatives that could be produced from bioconversion of saccharides. Some of these platform chemicals are already industrially produced via fermentation, such as bioethanol, citric acid, glutamic acid, lactic acid and 1,3-propanediol

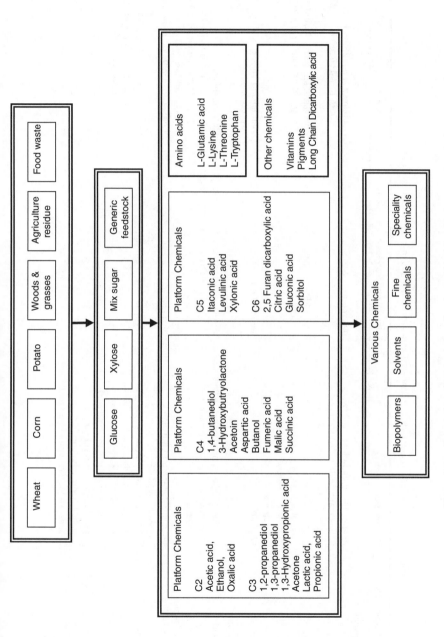

Figure 4.2 Generic biorefining schemes for chemical production from saccharides

Table 4.1 Platform chemicals that could be generated via bioconversion of saccharides and their potential derivatives

Platform chemicals	Possible derivatives	Potential applications of chemicals	References
C2 molecules			
Ethanol	Acetaldehyde, butadiene, diethyl succinate, ETBE (ethyl *tert*-butyl ether), ethyl acetate, ethylene	Fuel, monomers	Gucbilmez *et al.*, 2006; Patel *et al.*, 2006; Rass-Hansen *et al.*, 2007
Acetic acid	Acetic anhydride, chloroacetic acid, ethyl acetate, peracetic acid , vinyl acetate, cellulose acetate, terephthalic acid	Food additives, solvent, monomers, resin	Cheryan *et al.*, 1997; Ravinder *et al.*, 2000; Patel *et al.*, 2006
Oxalic acid	Oxalate	Fine chemicals	Mandal and Banerjee, 2005; Gadd, 1999
C3 molecules			
Lactic acid	2,3-Pentanedione, acrylic acid, acrylate polymers, cyclic lactide (dilactide), propylene glycol	Acidulant biopolymers, flavourings, pH buffers, preservatives, resins, solvent	Hofvendahl *et al.*, 1999; Hofvendahl and Hahn-Hagerdal, 2000; Datta and Henry, 2006
1,2-Propanediol		Polymers	Bennett and San, 2001
1,3-Propanediol	Polytrimethylene terephthalate, malonic acid	Polymers	Biebl *et al.*, 1999; Du *et al.*, 2007a
3-Hydroxy-propionic acid	1,3-Propanediol, acrylamide, acrylic acid, acrylonitrile, ethyl-3-hydroxypropionic acid, L-alanine, L-serine, malonic acid, propiolactone, poly(3-HP), poly(3-hydroxybutyric acid-co-3-hydroxypropionic acid)	Polymers, fine chemicals	Cao *et al.*, 1999; Werpy and Petersen, 2004; Zhang *et al.*, 2004; Patel *et al.*, 2006

Acetone		Biofuel, plastics, solvent	Ezeji et al., 2004; Ezeji et al., 2007b
Propionic acid		Cellulose plastics, preservatives, pesticides, perfumes, solvent	Lewis and Yang, 1992; Himmi et al., 1999
C4 molecules			
Butanol		Solvent, monomer	Qureshi et al., 2006; Ezeji et al., 2007a; Ezeji et al., 2007b
Butanediol	Polybutylene terephthalate (PBT), polybutylene succinate, polyurethanes, pyrrolidone, tetrahydrofuran, γ-butyrolactone	Biofuel, plastics, polymers, hot-melt adhesives	Biebl et al., 1998; Garg and Jain, 1995 ; Haas et al., 2005
Fumaric acid	1,4-Butanediol, L-alanine, L-aspartic acid, polyester resins, succinic acid, tetrahydrofuran, γ-butyrolactone	Unsaturated polyester resins, solvent, fine chemicals	Patel et al., 2006
Succinic acid	1,4-Butanediol, adipic acid, butyrate, diethyl succinate, dimethyl succinate, maleic anhydride, polyamides, polybutylene succinate, pyrrolidinones, succindiamide, tetrahydrofuran	Solvent, fine chemicals	Zeikus et al., 1999; Du et al., 2007b

(Continued)

Table 4.1 Platform chemicals that could be generated via bioconversion of saccharides and their potential derivatives (Continued)

Platform chemicals	Possible derivatives	Potential applications of chemicals	References
C5 molecules			
Itaconic acid	Itaconic anhydride	Artificial glass, resins fibres, lattices, detergents; bioactive compounds	Willke and Vorlop, 2001; Reddy and Singh, 2002; Levinson et al., 2006
Xylonic acid		Dispersant	Buchert et al., 1986; Buchert et al., 1988 Chun et al., 2006
C6 molecules			
Adipic acid	Nylon 66, esters, polyurethane resins	Fibres, resins, plasticisers, solvents, lubricants	Niu et al., 2002
Citric acid		pH Adjusters, chelating agents, antioxidants, preservatives	Grewal and Kalra, 1995; Roukas, 2000; Papagianni, 2007
Gluconic acid		Food additive	Roukas, 2000; Cheema et al., 2002; Singh et al., 2003; Erzinger and Vitolo, 2006
Sorbitol	1,4-Dorbitan, isosorbide, glycols, propylene glycol, polyetherpolyols, vitamin C	Surfactants, polymers, food additives	Ro and Kim, 1991; Cazetta et al., 2005; Erzinger and Vitolo, 2006
Glutamic acid	1,5-Pentanediol, 2-amino-1,5-pentanediol, 4-amino-5-hydroxypentanoic acid, 4-amino-butanoic acid, 5-methyl-2-pyrrolidinone, 5-hydroxynorvoline, 5-(hydroxymethyl)-2-pyrrolidinone, pyroglutamic acid, δ-caprolactone	Food additives, polymers, solvent	Hermann, 2003; Leuchtenberger et al., 2005; Corma et al., 2007

(Corn Refiners Association, 2006; Bevan and Franssen, 2006; Gray *et al.*, 2006). Lactic acid, for instance, which had a global market of 350 000 tonnes per year in 2000 (Corma *et al.*, 2007), is now predominantly produced via bacterial fermentation worldwide. Cargill operates the world's largest commercial lactic acid plant, which has a poly-lactic acid producing capacity of 140 000 tonnes per year, in Blair, Nebraska (Datta and Henry, 2006). On a lab scale, the highest lactic acid concentration attainable could reach 771 g L^{-1} by continuous extraction during fermentation, while the highest volumetric productivity could achieve 52–144 g L^{-1} h^{-1} when cell recycling was used (Hofvendahl and Hahn-Hagerdal, 2000). In industrial scale, it is believed that the lactic acid concentration, yield and productivity could reach 160–180 g L^{-1}, over 90% (w/w) and over 5 g L^{-1} h^{-1} (Patel *et al.*, 2006). Lactic acid has both a carboxylic group and a hydroxyl group, which makes it reactive in a variety of chemical conversions including polymerisation, esterification, dehydration, hydrogenation, reduction and oxidation, as shown in Figure 4.3.

Succinic acid is a potential platform chemical that is expected to be commercialised in a few years. Although the production capacity of petrochemically derived succinic acid is on the scale of 15 000 tonnes per year (Zeikus *et al.*, 1999), the production capacity of succinic acid derivatives is over 270 000 tonnes per year (Willke and Vorlop, 2004). Fermentative production of succinic acid could offer a viable route to bulk chemical production. Figure 4.4 presents potential routes for chemical production from succinic acid (McKinlay *et al.*, 2007). Another advantage of succinic acid microbial production is the simultaneous requirement for CO_2 consumption, which reduces the emission of the most important greenhouse gas and makes fermentative succinic acid production a process of significantly low environmental impact.

In comparison to lactic acid production, succinic acid fermentation results in low concentrations and productivities. Guettler *et al.* (1996) have reported the highest

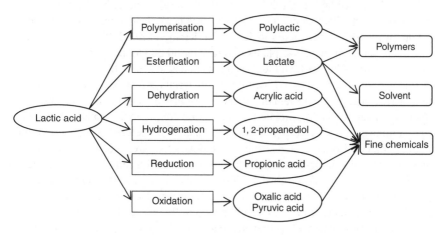

Figure 4.3 *Lactic acid as a building block for chemical production*

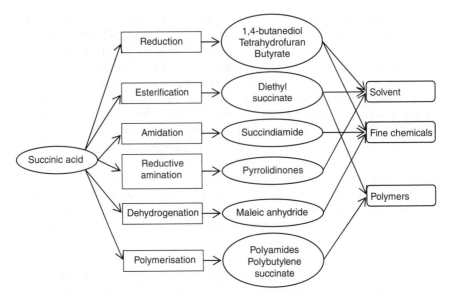

Figure 4.4 *Succinic acid as a building block for chemical production*

succinic acid concentration (up to 110 g L^{-1}) that has been achieved in *Actinobacillus succinogenes* cultivations on glucose-based media. The productivity achieved in this study was only in the range of 1.2–1.4 g L^{-1} h^{-1}. Meynial-Salles *et al.* (2007) reported the development of an integrated membrane-bioreactor-electrodialysis system, with which a 14.8 g L^{-1} h^{-1} succinic acid productivity could be achieved when operated at optimum conditions. Based on the current achievements, the overall succinic acid production cost will be around \$0.55–1.10 kg^{-1} (McKinlay *et al.*, 2007), which is close to the cost of the chemical production route. Two industrial scale succinic acid plants employing microbial production will be constructed in Japan (McKinlay *et al.*, 2007). The total production capacity of these two plants will be over 80 000 tonnes per year, which is around five times higher than the global succinic acid production.

Ethanol could be used as a substitute raw material for petroleum for the production of ethylene, propylene and butadiene (Klass, 1998). The main disadvantage is the low conversion yield as related to saccharide conversion, because ethanol has to be initially produced from glucose via fermentation before it is converted into building blocks. Both research groups and industrialists are focusing on improving and commercially implementing bioethanol conversion into ethylene (Varisli *et al.*, 2007; Chen *et al.*, 2007; Grubisich, 2007). Chen *et al.* (2007) reported ethylene production via catalytic dehydration of ethanol over TiO$_2$/γ-Al$_2$O$_3$ catalysts in multimicrochannel reactors with a conversion of 99.96% and ethylene selectivity of 99.4%. The ethylene yield achieved (26 g gcat^{-1} h^{-1}) could provide the basis for process intensification and miniaturisation of ethylene production from bioethanol (Chen *et al.*, 2007). The Brazilian company Braskem SA will launch a 200 000 metric tonnes per year polyethylene production plant, where the polymer would be produced from bioethanol-derived ethylene (Grubisich, 2007).

Fermentation has also been widely used in the production of many amino acids, such as glutamic acid, lysine, threonine and tryptophan (Hermann, 2003; Leuchtenberger *et al.*, 2005) with annual production of 1 500 000, 700 000, 30 000 and 1200 tonnes, respectively (Patel *et al.*, 2006). The amino acid concentration during fermentation could reach as high as 150 g L^{-1} glutamic acid, 100 g L^{-1} lysine, 100 g L^{-1} threonine, 58 g L^{-1} tryptophan, 51 g L^{-1} phenylalanine, 100 g L^{-1} arginine, 23 g L^{-1} histidine, 30 g L^{-1} isoleucine, 65 g L^{-1} serine, 100 g L^{-1} proline and 99 g L^{-1} valine (Demain, 2000; Leuchtenberger *et al.*, 2005). Apart from the traditional amino acid applications in food and animal feed, some amino acids could also be used as platform chemicals. Corma *et al.* (2007) reviewed the possible production routes using glutamic acid as a platform chemical for the production of various compounds. An interesting application is to produce polyglutamic acid (PGA), which is a non-toxic and biodegradable polymer with potential applications in medicine, food, cosmetic, plastics and water treatment (Corma *et al.*, 2007).

Besides fermentation, saccharides could also be subjected to enzymatic transformation or chemical conversion for bulk chemical production. For example, the acid-catalyzed cyclodehydration of pentose generates furfural, while the thermal acid-catalyzed dehydration of hexose leads to the production of 5-hydroxymethylfurfural (HMF) (Corma *et al.*, 2007). The latter could be further processed to levulinic acid. The oxidation of glucose results in glucaric acid, gluconic acid and 2-keto-gluconic acid. The reduction of hexose and pentose results in sorbitol, mannitol, xylitol and arabinitol. These chemicals could be used as solvents, monomers for polymer production and intermediates for the production of a large variety of derivatives. The biomass precursors and possible derivatives of some selected chemicals that could be derived from biomass are listed in Table 4.2. Among these, 2,5-furandicarboxylic acid, aspartic acid, glucaric acid, levulinic acid, 3-hydroxybutyrolactone and xylitol/arabinitol were selected among the top 12 potential platform chemicals by the US Department of Energy (Werpy and Petersen, 2004).

4.2.2 Chemical Production from Lignocellulosic Biomass

Lignocellulosic biomass is mainly constituted of cellulose, hemicellulose and lignin and is available in various forms, including agricultural residues, forestry residues, wood, energy crops and wastes from the paper, food and pulp industry. Conversion of such complex feedstocks into chemicals would require the development of commercially viable technologies that allow efficient and clean fractionation of lignocellulosic biomass into its components, followed by biological or chemical conversion, pyrolysis and gasification.

Table 4.2. Platform chemicals that could be generated via chemical conversion of saccharides and their possible derivatives

Platform chemicals	Raw material (biomass precursor)	Possible derivatives	References
2,5-Furandicarboxylic acid	Fructose, glucose (via HMF intermediate)	Diol, amine, levulinic acid, succinic acid, 2,5-furandicarbaldehyde, 2,5-dihydroxymethyl-furan, tetrahydrofuran, polyethylene, terephthalate analogues.	Werpy and Petersen 2004
3-Hydroxybutyrolactone	Starch	3-Hydroxytetrahydrofuran, 3-aminotetrahydrofuran, epoxy-lactone, furan, pyrrolidone, tetrahydrofuran	Werpy and Petersen 2004
5-Hydroxymethylfurfural	Fructose, glucose	2,5-bis(hydroxymethyl)-furan, 2,5-furandicarboxylic acid, 2,5-furandicarboxaldehyde, 5-fydroxymethyl-furanoic acid, levulinic acid	Bicker et al., 2003; Chheda et al., 2007
Aspartic acid	Fumaric acid	2-Amino-1,4-butanediol, 3-aminotetrahydrofuran, amino-2-pyrrolidone, aspartic anhydride	Werpy and Petersen 2004
Furfural	Pentose	Furfuryl alcohol, furfuryl amine, furan-acrylic acid, tetrahydrofuran, levulinic acid, maleic anhydride	Chheda et al., 2007; Corma et al., 2007
Glucaric acid	Starch, glucose	Lactone, polyglucaric ester, polyglucaric amide	Werpy and Petersen, 2004
Levulinic acid	Glucose, fructose, xylose	1,4-Pentanediol, 5-methyl-2-pyrrolidone, acrylic acid, diphenolic acid, methyl tetrahydofuran (MTHF), levulinate ester, succinic acid, β-acetyl-acrylic acid, γ-valero-lactone	Bozell et al., 2000; Keenan et al., 2004; Werpy and Petersen, 2004; Chang et al., 2007
Sorbitol	Glucose	1,4-Dorbitan, isosorbide, glycols, propylene glycol, polyetherpolyols, vitamin C	van Gorp et al., 1999; Castoldi et al., 2007

Integrated processes for fractionating the lignocellulosic matrix into its components, and hydrolysing cellulose and hemicellulose into the corresponding saccharides for chemical or biological conversion is not currently economically feasible. However, the enormous potential of chemical production from lignocellulosic biomass via integrated physical, chemical and biochemical processing has attracted research interest that could lead to viable processes in the mid-to-long term. Second-generation biorefineries utilising lignocellulosic biomass as raw material would mainly target the production of bioethanol through microbial fermentation of saccharides, predominantly glucose and xylose. However, the saccharides glucose and xylose, together with cellulose, hemicellulose and lignin, could be also used for chemical production.

The xylose produced via enzymatic hydrolysis of the hemicellulose fraction could be converted via microbial fermentation into xylitol (which has been listed by the US Department of Energy among the top 12 value-added platform chemicals) for the production of a range of chemicals, including xylaric acid, propylene glycol, ethylene glycol and a mixture of hydroxyl-furans and polyesters (Werpy and Petersen, 2004).

Cellulose could be selectively converted into various chemicals, including carboxylic acids, through catalytic sub-critical water oxidation using palladium or platinum catalysts (Schutta et al., 2002). Chang et al. (2007) also reported that wheat straw could be chemically converted into levulinic acid (another of the 12 building blocks) and further converted into gasoline and diesel oxygenates, biodegradable polymers, environmentally benign pesticides and solvents (Fitzpatrick, 2004). A novel thermochemical process called 'Biofine' has been presented that converts various sources of lignocellulosic biomass (e.g. bagasse, paper mill sludge) into useful platform chemicals, including levulinic acid, formic acid, furfural and a carbonaceous char suitable for conversion to synthesis gas (Fitzpatrick, 2004; Ritter, 2006). In the Biofine process, cellulose is converted into sugars via dilute sulfuric acid hydrolysis at 220 °C, which are subsequently transformed to levulinic acid. During the Biofine process, lignin is degraded into a carbon-rich char that is used to generate electricity. Levulinic acid is a precursor for methyltetrahydrofuran and ethyl levulinate (blends with diesel or gasoline create cleaner-burning fuels), δ-amino levulinic acid (used in herbicides, pesticides and cancer treatment) and diphenolic acid (used as alternative to bisphenol A in polycarbonates and phenolic resins).

Gravitis et al. (2001) also reported a novel technology consisting of a two-step selective catalysis of wood and other pentosan-containing raw materials for the production of various chemicals (i.e. furfural, levoglucosan, ethanol) by using small amounts of strong catalysts. Furfural could be used for the production of dyes, plastics and fumaric acid.

Apart from hemicellulose and cellulose, lignin is also a promising raw material for various applications. Huttermann et al. (2001) described various potential technologies for lignin modification for the production of new compounded materials. Pan et al. (2005) demonstrated that lignin extracted during ethanol organosolv pulping of softwood (the Lignol process) is a suitable feedstock for production of lignin-based adhesives and other products, due to its high purity, low molecular weight and abundance of reactive groups. Lignin-based fibres have attractive yields and can be readily stabilised, carbonised and graphitised for the production of carbon fibre composites to lower vehicle weight in order to decrease domestic vehicle fuel consumption (Griffith *et al.*, 2004). Potential chemicals from lignin are vanillin and complex mixtures of substituted quinines, phenols and catechols (Pollard, 2005).

Pyrolysis and gasification are promising technologies for the production of fuels and chemicals from lignocellulosic biomass (Demirbas, 2006). However, gasification of lignocellulosic biomass could lead to viable industrial processes in the short-term only, in the case of hydrogen and methanol production (Werpy and Peterson, 2004). Additional development of syngas production is required for the production of simple alcohols, aldehydes, mixed alcohols and Fischer–Tropsch liquids, as this stage accounts for at least 50% of the product cost and in many cases it is more like 75% (Spath and Dayton, 2003).

4.3   Vegetable Oils

Vegetable oils are mainly processed for chemical production either through hydrolysis or transesterification. Oil hydrolysis is carried out in pressurised water at 220 °C leading to the production of fatty acids and glycerol, whereas transesterification is a catalysed reaction leading to the production of fatty-acid alkyl esters and glycerol. Fatty acids, or in some cases oils, could be used for the production of surfactants (e.g. sorbitan ethoxylates, alkyl-polysaccharides), lubricants, dicarboxylic acids, resins, stabilisers, plasticisers, secondary alcohols and polyols (Pollard, 2005). Fatty-acid alkyl esters are mainly used as a biofuel, namely biodiesel (see Chapter 6 for more information). Biodiesel production involves transesterification of oils and/or fats with methanol and potassium or sodium hydroxide resulting in a methyl-ester biodiesel stream and a crude glycerol by-product stream that has few direct uses due to the presence of salts and other compounds.

According to the *European Directive for the Promotion of the Use of Biofuels* published by the European Council and the European Parliament in 2003, biodiesel production is expected to grow from around 6 million tonnes in the EU-25 in 2006 to 11 million tonnes in 2010. Increasing biodiesel production would proportionally result in the production of crude glycerol. The saturation of glycerol market is also a problem for various other companies that depend on revenues from glycerol sales (Willke and Vorlop, 2004; McCoy, 2005). Therefore, sustainable technologies converting the low-grade crude glycerol into bulk chemicals should be developed.

Considering the worldwide increase in biodiesel production, the market price of glycerol could be decreased from the current $1.34–2.00/kg to $0.45–1.12/kg making glycerol a major building block for bulk chemical production (Werpy and Petersen, 2004). Glycerol conversion could be accomplished by green chemical transformation, enzymatic or microbial bioconversion, and by processes integrating biological and chemical conversion. Research incentives have only been reported in the case of either biological or chemical conversion, whereas there are scarce literature-cited studies about processes combining these two technologies.

Glycerol could be transformed through green chemical processes into propylene glycol, propanol, branched polyesters, nylons, mono-, di- or tri-glycerate, diglyceraldehyde, glycerol carbonate and other oxidation products (Werpy and Petersen, 2004). Several research incentives focus on establishing, in the short term, commercially viable production of ethylene glycol and propylene glycol through hydrogenolysis of glycerol (Lahr and Shanks, 2003; Suppes *et al.*, 2007; Maris and Davis, 2007). Dow (and others) is currently developing a process to produce propylene glycol from glycerol (Anonymous, 2007). Huntsman Corp will scale up a process to make propylene glycol from glycerol at its process development facility in Conroe Texas by 2008 (Tullo, 2007). Agricultural processors Archer Daniels Midland and Cargill are also investigating propylene glycol production from glycerol (Tullo, 2007). Ethylene glycol and propylene glycol are produced in high quantities (ethylene glycol global annual demand is expected to reach 21 million tones by 2010) and have numerous applications, such as antifreeze compounds, de-icing compounds, and solvents.

Another important bulk chemical that could be derived from glycerol is acrylic acid (Craciun *et al.*, 2005; Shima and Takahashi, 2006; Dubois *et al.*, 2006). Shima and Takahashi (2006) reported a complete process for acrylic acid production involving the steps of glycerol dehydration in a gas phase followed by the application of a gas phase oxidation reaction to a gaseous reaction product formed by the dehydration reaction. Dehydration of glycerol could lead to commercially viable production of acrolein, which is an important and versatile intermediate for the production of acrylic acid esters, superabsorber polymers or detergents (Ott *et al.*, 2006). Sub- and supercritical water have been applied by Ott *et al.* (2006) as the reaction media for glycerol dehydration, but the conversion and acrolein selectivities that have been achieved so far are not satisfactory for an economical process.

Kraft *et al.* (2007) patented a process converting crude glycerol into chlorinated compounds, such as dichloropropanol and epichlorohydrin. Solvay and Dow are developing a process to convert glycerol to the epoxy resin raw material epichlorohydrin (Anonymous, 2007; Tullo, 2007).

Glycerol could be converted into hydrogen and alkanes through aqueous-phase reforming using a Pt/Al_2O_3 catalyst at temperatures and pressures around 500 K and 30 bars, respectively (Cortright *et al.*, 2002). Experimental studies illustrated that the selectivity for H_2 production improves in the order glucose < sorbitol < glycerol < ethylene glycol < methanol (Cortright *et al.*, 2002).

Several microorganisms could be adapted or engineered to consume glycerol as a carbon source for the production of chemicals that could be used either as end products or as important building blocks. Compared to saccharides, there are limited published studies on the utilisation of glycerol as a carbon source in microbial bioconversions. Glycerol as the sole carbon source has been mainly tested in preliminary studies for the production of succinic acid (Lee *et al.*, 2001; Dharmadi *et al.*, 2006), polyhydroxyalkanoates (Eggink *et al.*, 1994; Bormann and Roth, 1999; Ashby *et al.*, 2004; Koller *et al.*, 2005), 3-hydroxypropionaldehyde (Vancauwenberge *et al.*, 1990; Doleyres *et al.*, 2005), citric acid (Papanikolaou *et al.*, 2002), 3-hydroxypropionic acid (Suthers and Cameron, 2005), butanol (Biebl, 2001) and propionic acid (Barbirato *et al.*, 1997; Bories *et al.*, 2004). Glycerol conversion into 3-hydro- xypropionaldehyde (3HPA) could lead to a new platform molecule for the production of many chemicals (i.e. acrolein, acrylic acid, 1,3-propanediol) and polymers (Vollenweider and Lacroix, 2004). In contrast to other chemicals, extensive research has focused on the fermentative production of 1,3-propanediol from glycerol since there is so far no natural known microorganism producing this chemical from saccharides (Himmi *et al.*, 1999; Colin *et al.*, 2000; Papanikolaou *et al.*, 2000; Chen *et al.*, 2003; Hirschmann *et al.*, 2005). In fact, DuPont is currently developing a process to produce 1,3-propanediol from crude glycerol (Anonymous, 2007).

Glycerol could offer a significant advantage in several microbial fermentations as compared to glucose, because in certain cases it could lead to higher production yields and less by-product formation (Lee *et al.*, 2001; Bories *et al.*, 2004; Dharmadi *et al.*, 2006). However, intensive research is still required in order to develop bioprocessing schemes for viable chemical production from glycerol.

4.4   Chemical Production from Proteins

First generation bioethanol and biodiesel production, which mainly makes use of cereal grains and vegetable oils, represents a growing source of high quantities of protein as a valuable by-product. Sanders *et al.* (2007) estimated that a 10% substitution of fossil transportation fuels worldwide by first generation biofuels would result in an annual production of 100 million tonnes of protein – about four times the proteins requirement of the world's human population. A direct result of this would be the saturation of traditional protein markets. New opportunities would therefore emerge for chemical production from proteins.

Potential applications for plant proteins include polymers, coatings, composites, inks, cosmetics and personal care products, as well as two specialist applications in encapsulation materials and the production of vaccines and antibodies by genetically engineered plants (Pollard, 2005). Wheat gluten could be used for the production of biodegradable plastics, nanosized colloidal carriers and pharmaceuticals (Hernandez-Munoz *et al.*, 2003; Woerdeman *et al.*, 2004; Pommet *et al.*, 2005; Pallos *et al.*, 2006; Zhang *et al.*, 2006; Orecchioni *et al.*, 2006). Crude protein hydrolysates could also be used as fermentation supplements for the production of commodity and speciality chemicals (Franek *et al.*, 2000; Kwon *et al.*, 2005; Farges-Haddani *et al.*, 2006).

In the case that residual proteins from first generation biorefineries are efficiently hydrolysed into amino acids and these are separated, then they could be used for chemical production. For instance, wheat gluten is a rich source of glutamic acid and proline. Glutamic acid has been identified as a potentially very important platform molecule, which could be derivatised into a variety of chemicals, including polyglutamic acid, glutaric acid, 1,5-pentadiol, 5-amino-1-butanol, pyroglutamic acid, prolinol, norvoline and glutaminol (Werpy and Petersen, 2004; Holladay *et al.*, 2004). Aspartic acid is another potentially very interesting and important building block, which could be converted into 2-amino-1,4-butanediol, amino-2-pyrrolidone, aspartic anhydride, amino-γ-butyrolactone, 3-aminotetrahydrofuran and various substituted amino-diacids (Werpy and Petersen, 2004). Similarly, serine could be decarboxylated into ethanolamine and subsequently converted into 1,2-ethanediamine by the addition of ammonia (Sanders *et al.*, 2007). Arginine can be hydrolysed into ornithine, which could be converted into 1,4-butanediamine after decarboxylation (Sanders *et al.*, 2007).

4.5　Chemical Production Through Green Chemical Extraction of Biomass

Many types of biomass, including agricultural crops, energy crops and associated residues, may contain minor constituents with various useful applications. Green extraction processes, such as supercritical carbon dioxide, have been used widely in the past 25 years for the extraction of minor constituents, including secondary metabolites and oils. The application of supercritical fluid technology in herbal medicines (Lang and Wai, 2001), natural oils (Reverchon, 1997) and food additives (Herrero *et al.*, 2006) has been extensively studied. The feasibility of extracting valuable chemicals from agricultural residues, such as wheat straw, rye bran, rice bran and food waste has also been the focus of several research incentives (Francisco *et al.*, 2005; Deswarte *et al.*, 2006; Perretti *et al.*, 2003; Sabio *et al.*, 2003). The application of supercritical fluid techniques is now extended to the gasification and liquefaction of biomass for the production of biofuel or carbohydrate mixtures (Matsumura *et al.*, 2005; Matsumura *et al.*, 2006).

Protein-rich rape seed residues from biodiesel production plants are rich in phenolic compounds, glucocinolates and phytic acid. Phenolic compounds and phytic acid could be used as potent antioxidants in cosmetic and pharmaceutical formulations (Shamsuddin, 1995; Amarowicz and Shahidi, 1994; Wanasundara *et al.*, 1996; Oatway *et al.*, 2001). Derivatives from glucosinolate hydrolysis, including isothiocyanates, thiocyanates and nitriles, could be used as anticarcinogenic agents, biopesticides and flavour compounds (Halkier and Gershenzon, 2006).

Arabinoxylans is another group of compounds that could be extracted from biomass, and in particular wheat bran. Arabinoxylans is one of the major dietary fibres in wheat, which has numerous applications in bread manufacture, animal feeding and medical applications (Schooneveld-Bergmans, 1999). Hollmann and

Lindhauer (2005) recently carried out pilot-plant scale trials of an H_2O_2-based alkali extraction process, where a product was separated that had a purity of 70–80% arabinoxylans and a yield of 50% of the initial wheat bran. This process is expected to be commercialised as an integrated plant together with bioethanol production.

References

Amarowicz, R. and Shahidi, F. 1994. Chromatographic Separation of Glucopyranosyl Sinapate from Canola Meal. *J. American Oil Chem. Soc.*, **71**, 551–552.

Anonymous. June 2007. Process News: Cardiff Team to Turn Surplus Glycerol into Speciality Chemicals. *Chem. Eng.* Issue 792, 16.

Ashby, R.D., Solaiman, D.K.Y. and Foglia, T.A. 2004. Bacterial Poly(hydroxyalkanoate) Polymer Production from the Biodiesel Co-Product Stream. *J. Polym. Environ.*, **12**, 105–112.

Barbirato, F., Chedaille, D. and Bories, A. 1997. Propionic Acid Fermentation from Glycerol: Comparison with Conventional Substrates. *Appl. Microbiol. Biotechnol.*, **47**, 441–446.

Bennett, G.N. and San, K.Y. 2001. Microbial Formation, Biotechnological Production and Applications of 1,2-Propanediol. *Appl. Microbiol. Biotechnol.*, **55**, 1–9.

Bevan, M.W. and Franssen, M.C.R. 2006. Investing in Green and White Biotech. *Nat. Biotechnol.*, **24**, 765–767.

Bicker, M., Hirth, J. and Vogel, H. 2003. Dehydration of Fructose to 5-Hydroxymethylfurfural in Sub-and Supercritical Acetone. *Green Chem.*, **5**, 280–284.

Biebl, H., Zeng, A.P., Menzel, K. and Deckwer, W.D. 1998. Fermentation of Glycerol to 1,3-Propanediol and 2,3-Butanediol by *Klebsiella Pneumoniae. Appl. Microbiol. Biotechnol.*, **50**, 24–29.

Biebl, H., Menzel, K., Zeng, A.P. and Deckwer, W.D. 1999. Microbial Production of 1,3-Propanediol. *Appl. Microbiol. Biotechnol.*, **52**, 289–297.

Biebl, H. 2001. Fermentation of Glycerol by *Clostridium Pasteurianum* – Batch and Continuous Culture Studies. *J. Ind. Microbiol. Biotechnol.*, **27**, 18–26.

Bories, A., Himmi, E.H., Jauregui, J.J.A., Pelayo-Ortiz, C. and Gonzales, V.A. 2004. Glycerol Fermentation with *Propionibacteria* and Optimisation of the Production of Propionic Acid. *Sciences des Aliments,* **24**, 121–135.

Bormann, E.J. and Roth, M. 1999. The Production of Polyhydroxybutyrate by *Methylobacterium Rhodesianum* and *Ralstonia Eutropha* in Media Containing Glycerol and Casein Hydrolysates. *Biotechnol. Lett.,* **21**, 1059–1063.

Bozell, J.J., Moens, L., Elliott, D.C., Wang, Y., Neuenscwander, G.G., Fitzpatrick, S.W., Bilski, R.J. and Jarnefeld, J.L. 2000. Production of Levulinic Acid and Use as a Platform Chemical for Derived Products. *Resour. Conserv. Recy.,* **28**, 227–239.

Buchert, J., Viikari, L., Linko, M. and Markkanen, P. 1986. Production of Xylonic Acid by *Pseudomonas Fragi. Biotechnol. Lett.,* **8**, 541–546.

Buchert, J., Puls, J. and Poutanen, K. 1988. Comparison of *Pseudomonas Fragi* and *Gluconobacter Oxydans* for Production of Xylonic Acid from Hemicellulose Hydrolyzates. *Appl. Microbiol. Biotechnol.,* **28**, 367–372.

Cao, A., Arai, Y., Yoshie, N., Kasuya, K.I., Doi, Y. and Inoue, Y. 1999. Solid Structure and Biodegradation of the Compositionally Fractionated Poly(3-Hydroxybutyric Acid-Co-3-Hydroxypropionic Acid)s. *Polymer,* **40**, 6821–6830.

Cao, N. 1997. Production of Fumaric Acid by Immobilized *Rhizopus* Using Rotary Biofilm Contactor. *Appl. Biochem. Biotechnol. — Part A, Enz. Eng. Biotechnol.,* **63-65**, 387–394.

Castoldi, M., Camara, T., Monteiro, R.S., Constantino, A.M., Camacho, L., Carneiro, M. and Aranda, G. 2007. Experimental and Theoretical Studies on Glucose Hydrogenation to Produce Sorbitol. *React. Kinet. Catal. L,* **91**, 341–352.

Cazetta, M.L., Celligoi, M.A.P.C., Buzato, J.B., Scarmino, I.S. and da Silva, R.S.F. 2005. Optimization Study for Sorbitol Production by *Zymomonas Mobilis* in Sugar Cane Molasses. *Process Biochem.,* **40**, 747–751.

Chang, C., Cen, P. and Ma X. 2007. Levulinic Acid Production from Wheat Straw. *Biores. Technol.,* **98**, 1448–1453.

Cheema, J.J.S., Sankpal, N.V., Tambe, S.S. and Kulkarni, B.D. 2002. Genetic Programming Assisted Stochastic Optimization Strategies for Optimization of Glucose to Gluconic Acid Fermentation. *Biotechnol. Progr.,* **18**, 1356–1365.

Chen, G., Li, S., Jiao, F. and Yuan, Q. 2007. Catalytic Dehydration of Bioethanol to Ethylene Over TiO_2/γ-Al_2O_3 Catalysts in Microchannel Reactors. *Catal. Today,* **125**, 111–119.

Chen, X., Zhang, D.-J., Qi, W.-T., Gao, S.-.J, Xiu, Z.-L. and Xu, P. 2003. Microbial Fed-Batch Production of 1,3-Propanediol by *Klebsiella Pneumoniae* Under Micro-Aerobic Conditions. *Appl. Microbiol. Biotechnol.,* **63**,143–146.

Cheryan, M., Parekh, S., Shah, M. and Witjitra, K. 1997. Production of Acetic Acid by *Clostridium Thermoaceticum.Adv. Appl. Microbiol.,* **43**, 1–33.

Chheda, J.N., Román-Leshkov, Y. and Dumesic, J.A. 2007. Production of 5-Hydroxymethylfurfural and Furfural by Dehydration of Biomass-Derived Mono- and Poly-Saccharides. *Green Chem.,* **9**, 342–350.

Chun, B.-W., Dair, B., Macuch, P.J., Wiebe, D., Porteneuve, C. and Jeknavorian, A. 2006. The Development of Cement and Concrete Additive: Based on Xylonic Acid Derived via Bioconversion of Xylose.*Appl. Biochem. Biotechnol.,* **131**, 645–658.

Colin, T., Bories, A. and Moulin, G. 2000. Inhibition of *Clostridium Butyricum* by 1,3-Propanediol and Diols During Glycerol Fermentation. *Appl. Microbiol. Biotechnol.,* **54**, 201–205.

Corn Refiners Association. 2006. *Corn - Part of a Sustainable Environment.* Corn Refiners Association Annual Report 2006. http://www.corn.org/CRAR2006.pdf.

Corma, A., Iborra, S. and Velty, A. 2007. Chemical Routes for the Transformation of Biomass into Chemicals. *Chem. Rev.,* **107**, 2411–2502.

Cortright, R.D., Davda, R.R. and Dumesic, J.A. 2002. Hydrogen from Catalytic Reforming of Biomass-Derived Hydrocarbons in Liquid Water. *Nature,* **418**, 964–967.

Craciun, L., Benn, G.P., Dewing, J.R., Schriver, G.W., Peer, W.J., Siebenhaar, B. and Siegrist, U. 2005. *Preparation of Acrylic Acid Derivatives from Alpha or Beta-Hydroxy Carboxylic Acids.* Patent publication number WO2005095320.

Datta, R. and Henry, M. 2006. Lactic Acid: Recent Aadvances in Products, Processes and Technologies – A Review. *J. Chem. Technol. Biotechnol.,* **81**, 1119–1129.

Demain, A.L. 2000. Small Bugs, Big Business: The Economic Power of the Microbe. *Biotechnol. Adv.,* **18**, 499–514.

Demirbas, M.F. 2006. Hydrogen from Various Biomass Species via Pyrolysis and Steam Gasification Processes. *Energy Sources, Part A: Recov. Util. Environ. Effects,* **28**, 245–252.

Deswarte, F.E.I., Clark, J.H., Hardy, J.J.E. and Rose, P.M. 2006. The Fractionation of Valuable Wax Products from Wheat Straw Using CO_2. *Green Chem.,* **8**, 39–42.

Dharmadi, Y., Murarka, A. and Gonzalez, R. 2006. Anaerobic Fermentation of Glycerol by *Escherichia Coli*: A New Platform for Metabolic Engineering. *Biotechnol. Bioeng.,* **94**, 821–829.

Doleyres, Y., Beck, P., Vollenweider, S. and Lacroix, C. 2005. Production of 3-Hydroxypropionaldehyde Using a Two-Step Process with *Lactobacillus Reuteri*. *Appl. Microbiol. Biotechnol.*, **68**, 467–474.

Du, C., Zhang, Y., Li, Y. and Cao, Z. 2007a. Novel Redox Potential-Based Screening Strategy for Rapid Isolation of *Klebsiella Pneumoniae* Mutants with Enhanced 1,3-Propanediol-Producing Capability. *Appl. Environ. Microbiol.*, **73**, 4515–4521

Du, C., Lin, S., Koutinas, A., Wang, R. and Webb, C. 2007b. Succinic Acid Production from Wheat Using a Biorefining Strategy. *Appl. Microbiol. Biotechnol.*, DOI: 10.1007/s00253-007-1113-7

Dubois, J.-L., Duquenne, C. and Hoelderich, W. 2006. *Method for Producing Acrylic Acid from Glycerol*. Patent publication number WO2006114506.

Eggink, G., Eenink, A.H. and Huizing, H.J. 1994. *PHB-Producing Microorganism and Process for Removing Glycerol from a Culture Medium*. Patent publication number WO9409146.

Erzinger, G.S. and Vitolo, M. 2006. *Zymomonas Mobilis* as Catalyst for the Biotechnological Production of Sorbitol and Gluconic Acid. *Appl. Biochem. Biotechnol.*, **131**, 787–794.

Ezeji, T.C., Qureshi, N. and Blaschek, H.P. 2004. Acetone Butanol Ethanol (ABE) Production from Concentrated Substrate: Reduction in Substrate Inhibition by Fed-Batch Technique and Product Inhibition by Gas Stripping. *Appl. Microbiol. Biotechnol.*, **63**, 653–658.

Ezeji, T.C., Qureshi, N. and Blaschek, H.P. 2007a. Butanol Production from Agricultural Residues: Impact of Degradation Products on *Clostridium Beijerinckii* Growth and Butanol Fermentation. *Biotechnol. Bioeng.*, **97**, 1460–1469.

Ezeji, T.C., Qureshi, N. and Blaschek H.P. 2007b. Production of Acetone-Butanol-Ethanol (ABE) in a Continuous Flow Bioreactor Using Degermed Corn and *Clostridium Beijerinckii*. *Process Biochem.*, **42**, 34–39.

Farges-Haddani, B., Tessier, B., Chenu, S., Chevalot, I., Harscoat, C., Marc, I., Goergen, J.L. and Marc, A. 2006. Peptide Fractions of Rapeseed Hydrolysates as an Alternative to Animal Proteins in CHO Cell Culture Media. *Proc. Biochem.*, **41**, 2297–2304.

Fitzpatrick, S.W. 2004. Biofine Process: A Biorefinery Concept Based on Thermochemical Conversion of Biomass. *ACS National Meeting Book of Abstracts*, **227**, 173.

Francisco, J.D.C., Danielsson, B., Kozubek, A. and Dey, E.S. 2005. Application of Supercritical Carbon Dioxide for the Extraction of Alkylresorcinols from Rye Bran. *J. Supercrit. Fluids*, **35**, 220–226.

Franek, F., Hohenwarter, O. and Katinger, H. 2000. Plant Protein Hydrolysates: Preparation of Defined Peptide Fractions Promoting Growth and Production in Animal Cells Cultures. *Biotechnol. Progr.*, **16**, 688–692.

Gadd, G. 1999. Fungal Production of Citric and Oxalic Acid: Importance in Metal Speciation, Physiology and Biogeochemical Processes. *Adv. Microb. Physiol.*, **41**, 47–92.

Garg, S.K. and Jain, A. 1995. Fermentative Production of 2,3-Butanediol: A Review. *Bioresource Technol.*, **51**, 103–109.

Gravitis, J., Vedernikov, N., Zandersons, J. and Kokorevics, A. 2001. Furfural and Levoglucosan Production from Deciduous Wood and Agricultural Wastes. *ACS Symposium Series*, **784**, 110–122.

Gray, K.A., Zhao, L. and Emptage, M. 2006. Bioethanol. *Curr. Opin. Chem. Biol.*, **10**, 141–146.

Grewal, H.S. and Kalra, K.L. 1995. Fungal Production of Citric Acid. *Biotechnol. Adv.*, **13**, 209–234.

Griffith, W.L., Compere, A.L. and Leuten Jr., C.F. 2004. Lignin-Based Carbon Fiber for Transportation Applications. *2004 International SAMPE Technical Conference*, pp. 227–235.

Grubisich, J.C. 2007. Polyethylene from Sugarcane Ethanol in Brazil. *Ind. Bioproc.*, **29**, 8.

Gucbilmez, Y., Dogu, T. and Balci, S. 2006. Ethylene and Acetaldehyde Production by Selective Oxidation of Ethanol Using Mesoporous V-MCM-41 Catalysts. *Ind. Eng. Chem. Res.*, **45**, 3496–3502.

Guettler, M.V., Jain, M.K. and Rumler, D. 1996. *Method for Making Succinic Acid, Bacterial Variants for Use in the Process, and Methods for Obtaining Variants.* US Patent publication number 5573931.

Haas, T., Jaeger, B., Weber, R., Mitchell, S.F. and King, C.F. 2005. New Diol Processes: 1,3-Propanediol and 1,4-Butanediol. *Appl. Catal. A – Gen.*, **280**, 83–88.

Halkier, B.A. and Gershenzon, J. 2006. Biology and Biochemistry of Glucosinolates. *Annual Rev. Plant Biol.*, **57**, 303–333.

Hermann, T. 2003. Industrial Production of Amino Acids by Coryneform Bacteria. *J. Biotechnol.*, **104**,155–172.

Herrero, M., Cifuentes, A. and Ibanez, E. 2006. Sub- and Supercritical Fluid Extraction of Functional Ingredients from Different Natural Sources: Plants, Food-By-Products, Algae and Microalgae. *Food Chem.*, **98**, 136–148.

Hernandez-Munoz, P., Kanavouras, A., Perry, K.W.N. and Gavara, R. 2003. Development and Characterization of Biodegradable Films Made from Wheat Gluten Protein Fractions. *J. Agric. Food Chem.*, **51**, 7647–7654.

Himmi, E.H., Bories, A. and Barbirato, F. 1999. Nutrient Requirements for Glycerol Conversion to 1,3-Propanediol by *Clostridium Butyricum. Biores. Technol.*, **67**, 123–128.

Hirschmann, S., Baganz, K., Koschik, I. and Vorlop, K.-D. 2005. Development of an Integrated Bioconversion Process for the Production of 1,3-Propanediol from Raw Glycerol Waters. *Landbauforschung Volkenrode*, **55**, 261–267.

Hofvendahl, K., Akerberg, C., Zacchi, G. and Hahn-Hagerdal, B. 1999. Simultaneous Enzymatic Wheat Starch Saccharification and Fermentation to Lactic Acid by *Lactococcus Lactis. Appl. Microbiol. Biotechnol.*, **52**, 163–169.

Hofvendahl, K. and Hahn-Hagerdal, B. 2000. Factors Affecting the Fermentative Lactic Acid Production from Renewable Resources. *Enzyme Microb. Technol.*, **26**, 87–107.

Holladay, J.E., Werpy, T.A. and Muzatko, D.S. 2004. Catalytic Hydrogenation of Glutamic Acid. *Appl. Biochem. Biotechnol.*, **115**, 857–869.

Hollmann, J. and Lindhauer, M.G. 2005. Pilot-scale Isolation of Glucuronoarabinoxylans From Wheat Bran. *Carbohydr. Polym.*, **59**, 225–230.

Huttermann, A., Mai, C. and Kharazipour, A. 2001. Modification of Lignin for the Production of New Compounded Materials. *Appl. Microbiol. Biotechnol.*, **55**, 387–394.

Keenan, T.M., Tanenbaum, S.W., Stipanovic, A.J. and Nakas, J.P. 2004. Production and Characterization of Poly-β-Hydroxyalkanoate Copolymers from *Burkholderia Cepacia* Utilizing Xylose and Levulinic acid.*Biotechnol. Progr.*, **20**, 1697–1704.

Kim, H.S. and Ruy, D.D.Y. 1982. Continuous Glutamate Production Using an Immobilized Whole-Cell System. *Biotechnol. Bioeng.*, **24**, 2167–2174.

Kim, S. and Dale, B.E. 2004. Global Potential Bioethanol Production from Wasted Crops and Crop Residues. *Biomass Bioenerg.*, **26**, 361–375.

Klass, D.L. 1998. *Biomass for Renewable Energy, Fuels and Chemicals.* Academic Press, San Diego.

Koller, M., Bona, R., Braunegg, G., Hermann, C., Horvat, P., Kroutil, M., Martinz, J., Neto, J., Pereira, L. and Varila, P. 2005. Production of Polyhydroxyalkanoates from Agricultural Waste and Surplus Materials. *Biomacromol.*, **6**, 561–565.

Koutinas, A.A., Arifeen, N., Wang, R. and Webb, C. 2007c. Cereal-Based Biorefinery Development: Integrated Enzyme Production for Cereal Flour Hydrolysis. *Biotechnol. Bioeng.*, **97**, 61–92.

Koutinas, A.A., Malbranque, F., Wang, R.-H., Campbell, G.M. and Webb C. 2007b. Development of an Oat-Based Biorefinery for the Production of Lactic Acid by *Rhizopus Oryzae* and Various Value-Added Co-Products. *J. Agric. Food Chem.*, **55**, 1755–1761.

Koutinas, A.A., Wang, R.-H., Campbell, G.M. and Webb, C. 2006. A Whole Crop Biorefinery System: A Closed System for the Manufacture of Non-Food Products from Cereals, in *Biorefineries – Industrial Processes and Products*, Vol. 1, Kamm, B., Gruber, P.R. and Kamm, M. (Eds). Wiley-VCH, Weinheim, pp165–191.

Koutinas, A.A., Wang, R.-H. and Webb C. 2004a. Evaluation of Wheat as Generic Feedstock for Chemical Production. *Ind. Crops Prod.*, **20**, 75–88.

Koutinas, A.A., Wang, R.–H. and Webb, C. 2004b. Restructuring Upstream Bioprocessing: Technological and Economical Aspects for the Production of a Generic Microbial Feedstock from Wheat. *Biotechnol. Bioeng.*, **85**, 524–538.

Koutinas, A.A., Wang, R.-H. and Webb, C. 2007a. The Biochemurgist – Bioconversion of Agricultural Raw Materials for Chemical Production. *Biofuels Bioprod. Bioref.*, **1**, 24–38.

Koutinas, A.A., Xu, Y., Wang, R. and Webb, C. 2007d. Polyhydroxybutyrate Production from a Novel Feedstock Derived from a Wheat-Based Biorefinery. *Enzyme Microb. Technol.*, **40**, 1035–1044.

Kraft, P., Gilbeau, P., Gosselin, B. and Claessens, S. 2007. *Process for Producing Dichloropropanol from Glycerol, The Glycerol Coming Eventually from the Conversion of Animal Fats in the Manufacture of Biodiesel.* Patent publication number EP 1770081.

Kwon, M.S., Dojima, T. and Park, E.Y. 2005. Use of Plant-Derived Protein Hydrolysates for Enhancing Growth of *Bombyx Mori* (Silkworm) Insect Cells in Suspension Culture. *Biotechnol. Appl. Biochem.*, **42**, 1–7.

Lahr, D.G. and Shanks, B.H. 2003. Kinetic Analysis of the Hydrogenolysis of Lower Polyhydric Alcohols: Glycerol to Glycols. *Ind. Eng. Chem. Res.*, **42**, 5467–5472.

Lang, Q. and Wai, C.M. 2001. Supercritical Fluid Extraction in Herbal and Natural Product Studies – A Practical Review. *Talanta*, **53**, 771–782.

Lee, P.C., Lee, W.G., Lee, S.Y. and Chang, H.N. 2001. Succinic Acid Production with Reduced By-Product Formation in the Fermentation of *Anaerobiospirillum Succiniciproducens* Using Glycerol as a Carbon Source. *Biotechnol. Bioeng.*, **72**, 41–48.

Leuchtenberger, W., Huthmacher, K. and Drauz, K. 2005. Biotechnological Production of Amino Acids and Derivatives: Current Status and Prospects. *Appl. Microbiol. Biotechnol.*, **69**,1–8.

Levinson, W.E., Kurtzman, C.P. and Kuo, T.M. 2006. Production of Itaconic Acid by *Pseudozyma Antarctica* NRRL Y-7808 Under Nitrogen-Limited Growth Conditions. *Enzyme Microb. Technol.*, **39**, 824–827.

Lewis, V.P and Yang, S.T. 1992. Continuous Propionic Acid Fermentation by Immobilized *Propionibacterium Acidipropionici* in a Novel Packed-Bed Bioreactor. *Biotechnol. Bioeng.*, **40**, 465–474.

Mandal, S.K. and Banerjee, P.C. 2005. Submerged Production of Oxalic Acid from Glucose by Immobilized *Aspergillus Niger.Process Biochem.*, **40**, 1605–1610.

Maris, E.P. and Davis, R.J. 2007. Hydrogenolysis of Glycerol Over Carbon-Supported Ru and Pt Catalysts. *J. Catal.*, **249**, 328–337.

Matsumura, Y., Minowa, T., Potic, B., Kersten, S.R.A., Prins, W., van Swaaij, W.P.M., van de Beld, B., Elliott, D.C., Neuenschwander, G.G., Kruse, A. and Antal Jr, M.J. 2005. Biomass Gasification in Near- and Supercritical Water: Status and Prospects. *Biomass Bioenerg.*, **29**, 269–292.

Matsumura, Y., Sasaki, M., Okuda, K., Takami, S., Ohara, S., Umetsu, M. and Adschiri, T. 2006. Supercritical Water Treatment of Biomass for Energy and Material Recovery. *Combust. Sci. Technol.,* **178**, 509–536.

McCoy, M. 2005. An Unlikely Impact. *Chem. Eng. News,* **83**, 24–26.

McKinlay, J.B., Vieille, C. and Zeikus, J.G. 2007. Prospects for a Bio-Based Succinate Industry. *Appl. Microbiol. Biotechnol.,* **76**, 727–740.

Meynial-Salles, I., Dorotyn, S. and Soucaille, P. 2007. A New Process for the Continuous Production of Succinic Acid from Glucose at High Yield, Titer and Productivity. *Biotechnol. Bioeng.,* DOI, 10.1002/bit.21521.

Niu, W., Draths, K.M. and Frost, J.W. 2002. Benzene-Free Synthesis of Adipic Acid. *Biotechnol. Prog.,* **18**, 201–211.

Oatway, L., Vasanthan, T. and Helm, J.H. 2001. Phytic Acid. *Food Res. Internat.,* **17**, 419–431.

Orecchioni, A.-M., Duclairoir, C., Renard, D. and Nakache, E. 2006. Gliadin Characterization by Sans and Gliadin Nanoparticle Growth Modelization. *J. Nanosci. Nanotechnol.,* **6**, 3171–3178.

Ott, L., Bicker, M. and Vogel, H. 2006. Catalytic Dehydration of Glycerol in Sub- and Supercritical Water: A New Chemical Process for Acrolein Production. *Green Chem.,* **8**, 214–220.

Pallos, F.M., Robertson, G.H., Pavlath, A.E. and Orts, W.J. 2006. Thermoformed Wheat Gluten Biopolymers. *J. Agric. Food Chem.,* **54**, 349–352.

Pan, X., Arato, C., Gilkes, N., Gregg, D., Mabee, W., Pye, K., Xiao, Z., Zhang, X. and Saddler, J. 2005. Biorefining of Softwoods Using Ethanol Organosolv Pulping: Preliminary Evaluation of Process Streams for Manufacture of Fuel-Grade Ethanol and Co-Products. *Biotechnol. Bioeng.,* **90**, 473–481.

Papagianni, M. 2007. Advances in Citric Acid Fermentation by *Aspergillus Niger*: Biochemical Aspects, Membrane Transport and Modeling. *Biotechnol. Adv.,* **25**, 244–263.

Papanikolaou, S., Muniglia, L., Chevalot, I., Aggelis, G. and Marc, I. 2002. *Yarrowia Lipolytica* as a Potential Producer of Citric Acid from Raw Glycerol. *J. Appl. Microbiol.,* **92**, 737–744.

Papanikolaou, S., Ruiz-Sanchez, P., Pariset, B., Blanchard, F. and Fick, M. 2000. High Production of 1,3-Propanediol from Industrial Glycerol by a Newly Isolated *Clostridium Butyricum* Strain. *J. Biotechnol.,* **77**, 191–208.

Patel, M., Crank, M., Dornburg, V., Hermann, B., Roes, L., Husing, B., Overbeek, L., Terragni, F. and Recchia, E. 2006. The BREW Project Report: Medium and Long-Term Opportunities and Risks of the Biotechnological Production of Bulk Chemicals from Renewable Resources. http://www.chem.uu.nl/brew/BREW_Final_Report_September_2006.pdf

Perretti, G., Miniati, E., Montanari, L. and Fantozzi, P. 2003. Improving the Value of Rice By-Products by SFE. *J. Supercrit. Fluids,* **6**, 63–71.

Pollard, G. 2005. *Catalysis in Renewable Feedstocks – A Technology Roadmap.* Report prepared on behalf of The Department of Trade and Industry, Report number CR7656. BHR Solutions Project No 180 2421. http://www.bhrgroup.co. uk/extras/renewcatfull.pdf

Pommet, M., Redl, A., Guilbert, S. and Morel, M.-H. 2005. Intrinsic Influence of Various Plasticizers on Functional Properties and Reactivity of Wheat Gluten Thermoplastic Materials. *J. Cereal Sci.,* **42**, 81–91.

Qureshi, N., Li, X.L., Hughes, S., Saha, B.C. and Cotta, M.A. 2006. Butanol Production from Corn Fiber Xylan Using *Clostridium Acetobutylicum. Biotechnol. Prog.,* **22**, 673–680.

Rass-Hansen, J., Falsig, H., Joergensen, B. and Christensen, C.H. 2007. Bioethanol: Fuel or Feedstock? *J. Chem. Technol. Biotechnol.,* **82**, 329–333.

Ravinder, T., Ramesh, B., Seenayya, G. and Reddy, G. 2000. Fermentative Production of Acetic Acid from Various Pure and Natural Cellulosic Materials by *Clostridium Lentocellum* SG6. *W. J. Microbiol. Biotechnol.,* **16**, 507–512.

Reddy, C.S.K. and Singh, R.P. 2002. Enhanced Production of Itaconic Acid from Corn Starch and Market Refuse Fruits by Genetically Manipulated *Aspergillus Terreus* SKR10. *Bioresource Technol.*, **85**, 69–71.

Reverchon, E. 1997. Supercritical Fluid Extraction and Fractionation of Essential Oils and Related Products. *J. Supercrit. Fluids,* **10**, 1–37.

Ritter, S. 2006. Biorefinery Gets Ready to Deliver the Goods. *Chem. Eng. News.*, **84**(34), 47.

Ro, H.S. and Kim, H.S. 1991. Continuous Production of Gluconic Acid and Sorbitol from Sucrose Using Invertase and an Oxidoreductase of *Zymomonas Mobilis. Enzyme Microb. Technol.*, **13**, 920–924.

Roukas, T. 2000. Citric and Gluconic Acid Production from Fig by *Aspergillus niger* Using Solid-State Fermentation. *J. Ind. Microbiol. Biotechnol.*, **25**, 298–304.

Sabio, E., Lozano, M., Montero de Espinosa, V., Mendes, R.L., Pereira, A.P., Palavra, A.F. and Coelho, J.A. 2003. Lycopene and β-Carotene Extraction from Tomato Processing Waste Using Supercritical CO_2. *Ind. Eng. Chem. Res.*, **42**, 6641–6646.

Sanders, J., Scott, E., Weusthuis, R. and Mooibroek, H. 2007. Bio-Refinery as the Bio-Inspired Process to Bulk Chemicals. *Macromol. Biosci.*, **7**, 105–117.

Schooneveld-Bergmans, M.E.F., Van Dijk, Y.M., Beldman, G. and Voragen, A.G.J. 1999. Physicochemical Characteristics of Wheat Bran Glucuronoarabinoxylans. *J. Cereal Sci.*, **29**, 49–61.

Schutta, B.D., Serrano, B., Cerro, R.L. and Abraham, M.A. 2002. Production of Chemicals from Cellulose and Biomass-Derived Compounds Through Catalytic Sub-Critical Water Oxidation in a Monolith Reactor. *Biom. Bioen.*, **22**, 365–375.

Shamsuddin, A.M. 1995. Inositol Phosphates Have Novel Anticancer Function. *J. Nutrition,* **125**, 725S–732S.

Shima, M. and Takahashi, T. 2006. *Method for Producing Acrylic Acid*. Patent publication number EP1710227.

Singh, O.V., Jain, R.K. and Singh, R.P. 2003. Gluconic Acid Production Under Varying Fermentation Conditions by *Aspergillus Niger. J. Chem. Technol. Biotechnol.*, **78**, 208–212.

Spath, P.L. and Dayton, D.C. 2003. *Preliminary Screening – Technical and Economic Assessment of Synthesis Gas to Fuels and Chemicals with Emphasis on the Potential for Biomass-Derived Syngas*. National Renewable Energy Laboratory, NREL/TP-510-34929, http://www.nrel.gov/docs/fy04osti/34929.pdf

Suppes, G.J., Sutterlin, W.R. and Dasari, M.A. 2007. *Method of Producing Lower Alcohols from Glycerol*. Patent publication number EP1727875.

Suthers, P.F. and Cameron, D.C. 2005. *Production of 3-Hydroxypropionic Acid in Recombinant Organisms*. United States Patent number 6852517.

Taing, O. and Taing, K. 2007. Production of Malic and Succinic Acids by Sugar-Tolerant Yeast *Zygosaccharomyces Rouxii. Eur. Food Res. Technol.*, **224**, 343–347.

Tullo, A. 2007. News of the Week: Chemicals from Renewables. *Chem. Eng. News,* **85**(19), 14.

van Gorp, K., Boerman, E., Cavenaghi, C.V. and Berben, P.H. 1999. Catalytic Hydrogenation of Fine Chemicals: Sorbitol Production. *Catal. Today,* **52**, 349–361.

Vancauwenberge, J.E., Slininger, P.J. and Bothast, R.J. 1990. Bacterial Conversion of Glycerol to 3-Hydroxypropionaldehyde. *Appl. Environ. Microbiol.*, **56**, 329–332.

Varisli, D., Dogu, T. and Dogu, G. 2007. Ethylene and Diethyl-Ether Production by Dehydration Reaction of Ethanol Over Different Heteropolyacid Catalysts. *Chem. Eng. Sci.*, **62**, 5349–5352.

Vollenweider, S. and Lacroix, C. 2004. 3-Hydroxypropionaldehyde: Applications and Perspectives of Biotechnological Production. *Appl. Microbiol. Biotechnol.*, **64**, 16–27.

Wanasundara, U.N., Amarowicz, R. and Shahidi, F. 1996. Partial Characterization of Natural Antioxidants in Canola Meal. *Food Res. Internat.,* **28**, 525–530.

Webb, C., Koutinas, A.A. and Wang, R. 2004. Developing a Sustainable Bioprocessing Strategy Based on a Generic Feedstock. *Adv. Biochem. Eng./Biot.,* **87**, 195–268.

Werpy, T. and Petersen, G. 2004. *Top Value Added Chemicals from Biomass. Volume I – Results of Screening for Potential Candidates from Sugars and Synthesis Gas.* Produced by staff at the Pacific Northwest National Laboratory (PNNL) and the National Renewable Energy Laboratory (NREL). http://www1.eere.energy.gov/biomass/pdfs/ 35523.pdf.

Wilke, D. 1999. Chemicals From Biotechnology: Molecular Plant Genetics Will Challenge the Chemical and the Fermentation Industry. *Appl. Microbiol. Biotechnol.,* **52**, 135–145.

Willke, T. and Vorlop, K.D. 2001. Biotechnological Production of Itaconic Acid. *Appl. Microbiol. Biotechnol.,* **56**, 289–295

Willke, T. and Vorlop, K.D. 2004. Industrial Bioconversion of Renewable Resources as an Alternative Chemistry. *Appl. Microbiol. Biotechnol.,* **66**, 131–142.

Woerdeman, D.L., Veraverbeke, W.S., Parnas, R.S., Johnson, D., Delcour, J.A. and Verpoest, I. 2004. Designing New Materials from Wheat Protein. *Biomacromol.,* **5**, 1262–1269.

Yahiro, K., Shibata, S., Jia, S.-R., Park, Y. and Okabe, M. 1997. Efficient Itaconic Acid Production from Raw Corn Starch. *J. Ferm. Bioeng.,* **84**, 375–377.

Zeikus, J.G., Jain, M.K. and Elankovan, P. 1999. Biotechnology of Succinic Acid Production and Markets for Derived Industrial Products. *Appl. Microbiol. Biotechnol.,* **51**, 545–552.

Zhang, D., Hillmyer, M.A. and Tolman, W.B. 2004. A New Synthetic Route to Poly[3-Hydroxypropionic Acid] (P[3-HP]): Ring-Opening Polymerization of 3-HP Macrocyclic Esters. *Macromol.,* **37**, 8198–8200.

Zhang, X., Do, M.D., Hoobin, P. and Burgar, I. 2006. The Phase Composition and Molecular Motions of Plasticized Wheat Gluten-Based Biodegradable Polymer Materials Studied by Solid-State NMR Spectroscopy. *Polymer,* **47**, 5888–5896.

5

Biomaterials

Carlos Vaca-Garcia

Institut National Polytechnique de Toulouse France (INP-ENSIACET)

5.1 Introduction

Man-made materials are obtained from one of these three resources: mineral (e.g. silica), fossil (e.g. petrol, natural gas) and vegetable (e.g. cotton, wood). The materials obtained from each of them are usually exclusive to their resource (glass can only be made out of mineral resources and cellophane can only be obtained starting with vegetable resources). Sometimes, the same material can be obtained from fossil or vegetable resources if more than one transformation step is performed. For instance, polyethylene is the product of the polymerisation of ethylene. This gas can be obtained from the petrochemical industry by steam cracking of light petrol molecules. Surprisingly, it can also be obtained from the gasification of cellulose followed by the Fischer–Tropsch process. The choice of both the source and the process then depends on the availability and price of the resource and the feasibility and economy of the process.

A second approach to better understand the development of new materials (biomaterials for instance) is the following. If a present material becomes scarce or too expensive, another material, more abundant and/or cheaper, can be used provided it fulfils the basic requirements of performance. For instance, accessory metal parts in cars were replaced by reinforced plastic because of the lower price and lower weight during the period of a steel crisis.

Introduction to Chemicals from Biomass Edited by James Clark and Fabien Deswarte
© 2008 John Wiley & Sons, Ltd

A third approach to be considered is the ecological and sustainable aspects of a material. For instance, the fossil resources used for the fabrication of plastics constitute an undesirable contribution of carbon dioxide to the atmosphere thus increasing global warming. Moreover, the long persistence of some plastic materials in the environment causes visual pollution and suffocation of animals, if their disposal in the post-use period is not adequate. The energy consumption in the fabrication of a material also falls under this consideration.

Therefore, the choice of a material and its source has become a new challenge in our modern world. There is no reliable warranty on the supply of raw materials on our planet and we grow more and more ecologically conscious. A replacement policy has already started and the benefit goes to biomaterials. In this chapter we will define biomaterials as those materials obtained from renewable resources. These biological sources can grow fast. Plants naturally convert carbon dioxide from the atmosphere into polymers (so-called biopolymers) and other compounds, such as sugars and lipids, are easily convertible into materials. Biomaterials are therefore the best choice to regulate the carbon cycle in the lithosphere, provided that their life cycle analysis is positive, which is not a general rule.

5.2   Wood and Natural Fibres

Nature has given mankind a large palette of biomaterials from plants. The fact that they are natural does not mean that their performances are poor, as might be wrongly thought. Wood, for instance, combines elasticity, insulation and toughness. Linen provides a fresh and elegant textile in summer. Stubble in thatched roofs is still used in rural houses or even in ancient Japanese temples; its strength and insulation properties account for this preference.

The common elements in the cited examples are the mechanical characteristics and the insulation properties. These advantages come, not only from macroscopic configuration of these materials (the hollow cylindrical structure of stubble for instance), but, mainly, from their microscopic structure. Most the vegetal fibres can be described by two models: wood fibres and cotton fibres, which will be presented later. In order to better understand the mechanical properties of these fibres, let us first consider their molecular constitution, then their hierarchical structure.

5.2.1   Molecular Constitution

All natural fibres are constituted essentially of four kinds of components:

- Structural polysaccharides: cellulose and hemicelluloses (e.g. xylan)
- Other polysaccharides (e.g. pectin)
- Lignin
- Water- and solvent-soluble compounds (e.g. waxes, minerals).

Cellulose is by far the most abundant component, especially in cotton fibres. This biopolymer is constituted of cellobiose (repetitive unit), which is composed

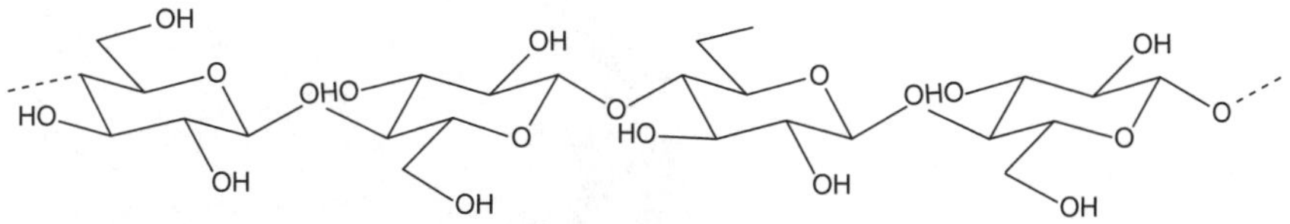

Figure 5.1 *Molecule of cellulose*

of two glucose building blocks in reverse positions with respect to the plane of the cycle (Figure 5.1).

Such configuration results from a β-1 $\rightarrow$ 4 glucosidic bond. Schematically speaking, cellulose is a 'flat' molecule with its hydroxyl groups pointing out of the 'ribbon'. The accessibility of OH groups favours the formation of intermolecular hydrogen bonds and the creation of crystalline zones. A cellulose molecule from a cotton fibre contains more than 7000 glucose units. After extraction, the degree of polymerisation is reduced at about 1000 units.

Hemicelluloses are constituted of different hexoses and pentoses: glucose, mannose, xylose, etc. Since these heteropolysaccharides are often branched polymers, they cannot constitute crystalline structures. However, their function in the constitution of natural fibres is crucial. Together with lignin, they constitute the bonding matrix of the cellulose microfibres.

Lignin is an amorphous non-polar macromolecule, constituted of phenylpropane units. The structure of lignin depends on the source. Moreover, the extraction method modifies the structure of the lignin prior to analysis. Recent studies (Banoub *et al.*, 2007; Forss and Fremer, 2000) have shown fundamental differences with the known structures of lignin.

In the classic structures, three basic units can be found (Figure 5.2):

- The guaiacyl unit, derived from the *trans*-coniferyl alcohol, abundantly present in softwoods
- The syringyl unit, derived from the *trans*-sinapyl alcohol, which is present with the guaiacyl unit in hardwoods

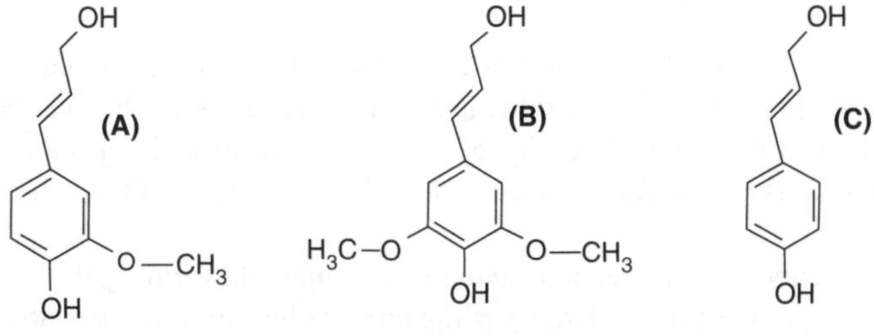

Figure 5.2 *Molecules from which the characteristics units of lignin are obtained: trans-coniferyl alcohol (A), trans-sinapyl alcohol (B) and trans-p-cumaryl alcohol (C)*

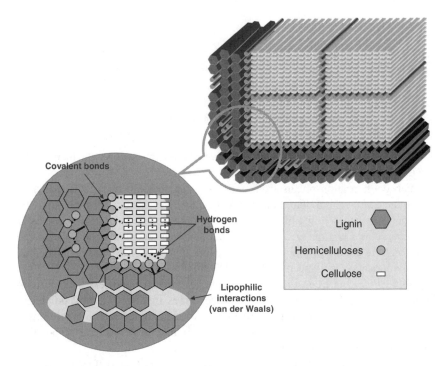

Figure 5.3 *Schematic representation of the lignin-carbohydrate complex (LCC) and its interaction with cellulose*

- The *p*-hydroxyphenyl unit, derived from the *trans-p*-cumaryl alcohol, characteristic of grasses.

Lignin is linked through covalent bonds (ester and ether) to hemicelluloses. The two macromolecules then constitute the lignin–carbohydrate complex (known as LCC). As the hemicelluloses can be linked to cellulose through hydrogen bonding, the LCC is capable to assemble the cellulose microfibrils (Figure 5.3).

5.2.2   Wood and Timber

Employed for thousands of years as a structural material or source of energy, wood is still a preferred construction material because of its mechanical properties and its visual aspect. The diversity of wood species throughout the world and the elevated world production makes an abundant material, available in almost all countries.

Wood is defined as the ligneous and compact material forming the branch, the trunk and the roots of trees. Timber is the term dedicated to sawn wood, usually the trunk of the tree, used for construction purposes due to its strength. Timber production has risen to 3340 million m^3 worldwide. In Europe, Sweden, Finland

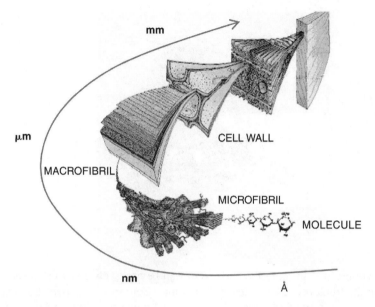

mm

μm

CELL WALL

MACROFIBRIL

MICROFIBRIL

MOLECULE

nm

Å

Figure 5.4   *Hierarchical structure of wood: from timber to cellulose*

and Germany produce almost half of the EU-25 production, which has risen to 367.4 million m^3 (source: FAO).

From the nano scale to the macroscopic aspect, wood is structured in a hierarchical way. The molecules of cellulose (40–50%) are arranged first in microfibrils, then in macrofibrils and finally they constitute the layers of the wall of every cell in the wood (Figure 5.4).

The cell wall (Figure 5.5) is composed of a primary and a secondary wall surrounding a void, the lumen (L). The primary wall (0.1–0.2 μm) presents no particular arrangement and is constituted of cellulose, hemicelluloses, pectin, proteins and lignin. The secondary wall consists of three layers namely S1, S2 and S3, all of which are oriented layers of cellulose:

• In the S1 layer (0.2–0.3 μm) and in the S3 layer (0.1 μm), the microfibrils make an angle of 50 to 70° to the axis of the fibre.
• In the S2 layer (1–5 μm), which constitutes 90% of the weight of the cell wall, the angle of the microfibrils is between 10 and 30°. The shorter this angle, the higher the modulus and the strength of the fibre (and the lower the elongation at break).

Two neighbouring cells are separated by the middle lamella (M). It is constituted of polysaccharides and lignin, as explained in Figure 5.3. Its thickness is between 0.2 and 1 μm.

At a mesoscopic level, the hollow cylindrical-shaped cells are orientated in the same direction as the trunk. Radial sheets of horizontal cells rich in lignin go from

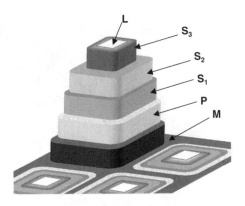

Figure 5.5   *Schematic projection of a wood cell showing its different layers*

the centre of the trunk to the bark. They can be easily observed in a transversal cut of the trunk because of their darker colour. Both the cells orientation and the rays cause anisotropy in wood. The mechanical properties of wood depend on the direction of the mechanical solicitation (Figure 5.6).

Let us consider a wooden pencil; it is quite easy to crush it with our teeth in a horizontal position whereas it is quite difficult to do the same if we place the pencil in a vertical position (let us imagine a reasonably short pencil and good teeth!). A Canadian lumberjack always cuts the small wood rods with his axe in a vertical (axial) position. Even the name of the tool seems to give him reason! The explanation is found in Figure 5.7.

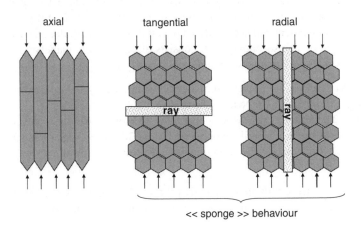

Figure 5.6   *Schematic representation of the three different compression solicitations in wood according to the position of the cells and the ray*

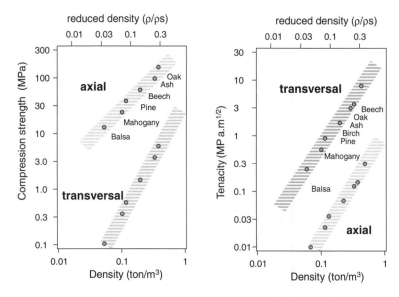

Figure 5.7   Mechanical properties of various wood species as a function of wood density

In other words, the pencil is between 30 to 100 times easier to crunch in a transversal position depending on the wood species. A similar analysis shows that the lumberjack requires about 30 times less effort to cut the rods in a vertical direction. Table 5.1 shows the main mechanical properties of different wood species.

The water content in wood influences its mechanical properties, but also changes the dimensions. When completely dried wood is placed in a humid atmosphere, water uptake makes the cells swell in a linear relation with the water content of the wood. When the fibre reaches the saturation point (around 30% water according to species), the water uptake can continue (especially if wood is in contact with liquid water), but the swelling stops. Swelling is different according to the section of the wood. Axial maximum swelling is limited to less than 1%. The transversal swelling is more significant: radial 4% and tangential 7% approximately. Swelling is a reversible phenomenon: humid wood shrinks during drying. Hysteresis is always observed. For some applications, the dimensional instability of wood is a serious problem. Hydrophobic agents can be added, but chemical modification is also used. For instance, acylation of wood with aliphatic anhydrides limits the swelling to 75% (Hill and Jones, 1996).

Despite all the advantages of wood, its use as a material is conditioned to its durability. Softwoods have a limited durability and hardwoods are, in general, more resistant than softwoods, but both contain excellent nutriments for fungi and xylophagous insects (termites, longhorn beetle larvae, etc.). The impregnation of wood with antifungal and insecticide compounds is a current practice. Among

Table 5.1 Mechanical properties of selected wood species with a water content of 12%

Wood species	Density (ton·m⁻³)	Fibre length (mm)	Strength at break (MPa)			Flexural modulus (GPa)
			Compression	Tensile	Flexural	
Softwoods						
Spruce, fir, red cedar	0.3–0.9	3–7	35–45 axial 6–8 ⊥	90–100 axial 1.2 ⊥	50–70	8–10
Pines	0.3–0.85	4	40–50 axial 7.5–8 ⊥	100–120 axial 1.8 ⊥	80–90	9–14
Poplar, mahogany, okoume	0.3–0.5	1.1	30–40 axial 7.5–10 ⊥	80–100 axial 2 ⊥	65–85	9–11
Hardwoods						
Chestnut, soft oak, maple, ash, beech	0.3–0.95	1.1–1.2	40–60 axial 12–15 ⊥	100–120 axial 3 ⊥	75–130	9–12.5
Hard oak, teck	0.3–0.8	0.7–1.2	50–80 axial 18–20 ⊥	120–150 axial 4 ⊥	100–170	11.5–15
Azobe	1.1		90–100 axial >20 ⊥	150–200 axial 5 ⊥	227	

Sources: www.matweb.com and Dulbecco and Luro (2001).

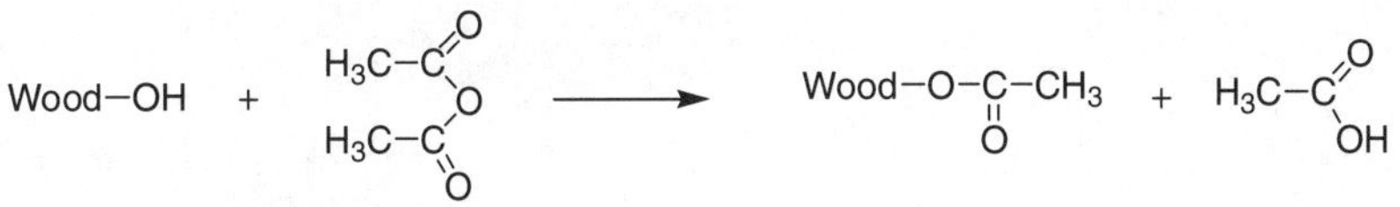

Figure 5.8   *Reaction of wood acetylation*

them, soluble salts of copper, chromium and arsenic (CCA) are the most used. Human health concerns have forced wood treatment units to eliminate the most toxic compounds in certain countries. Thus, CCB (B stands for boron), or CCO (without boron) have appeared, for instance. In Europe, thanks to the directive *Biocides* 98/8/CE, wood treatment chemicals will be regulated from 2008 to ensure that they do not present a significant danger to the environment, to humans or to animals.

Following these ecological concerns, new industrial ways to preserve wood have been developed from intensive laboratory and pilot-plant research. Among them we can note:

- Acetylation [e.g. Accoya (www.accoya.info)]: Wood is treated with acetic anhydride. Residual acetic acid from the esterification reaction is difficult to eliminate. Residual vinegar odour may appear (Figure 5.8).
- Thermal treatment [e.g. Retiwood (www.retiwood.com), WTT (www.wtt.dk)]: Wood undergoes partial pyrolysis at high temperature in the absence of oxygen or in the presence of steam or reducing compounds. Rotting biomolecules are thus destroyed.
- Oleothermal treatment [e.g. Oléobois (Dumonceaud and Thomas, 2004)]. Wood is 'fried' in hot vegetable oils, which make a barrier to aggressors, especially by the cross-linking of triglycerides. The use of siccative oils (linseed, sunflower) increases the efficacy of the treatment.
- Oleochemical modification [e.g. Surfasam (Morard *et al.*, 2007), WoodProtect (Magne *et al.*, 2003)]. Wood is reacted with fatty compounds such as alkenyl succinic anhydrides or fatty-acetic anhydrides. The alkenyl or fatty base comes from vegetable oils (fatty acid or fatty acid esters) (Figure 5.9).

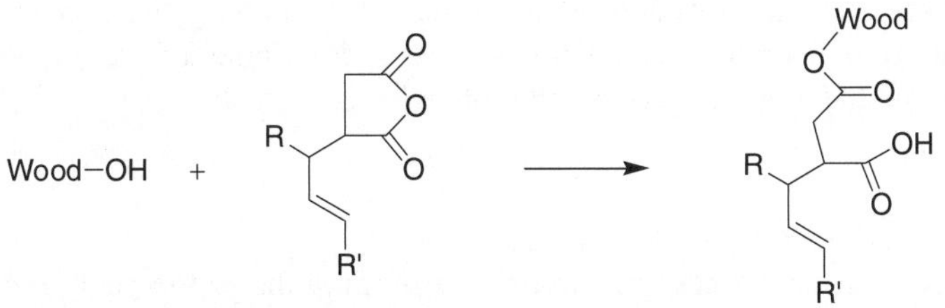

Figure 5.9   *Reaction of wood with alkenyl succinic anhydrides (as in Surfasam treatment)*

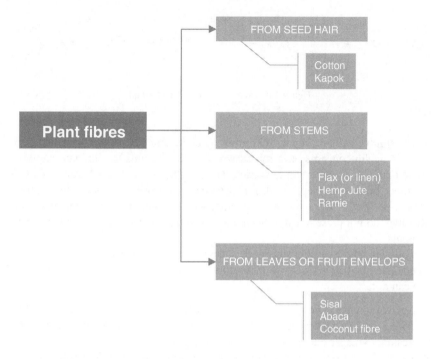

Figure 5.10   *Plant fibres classified according to their position in the plant*

5.2.3   Plant Fibres

Plant fibres are as diversified as wood species. Their properties are variable according to their species, their age, their growing position, etc. They can be classified in three categories (Figure 5.10).

By far the most produced fibre in the world is cotton. Its production is almost 30 times that of jute (Figure 5.11). The extended industrial textile applications of cotton account for such huge production. Nevertheless, it is the other fibres that occupy an important place in biomaterials production. We should also note that most of the natural fibres (except wood) come from emerging countries, especially from Asia.

The fibre characteristics depend on the source. Table 5.2 gathers the main characteristics of industrial vegetable fibres. A short description of every type of fibre and their applications are given in the following paragraphs.

Cotton

Cotton is a soft fibre that grows around the seeds of the cotton plant (*Gossypium spp.*), but practically all of the commercial cotton grown worldwide today comes from the American species *Gossypium hirsutum* and *Gossypium barbadense*. The

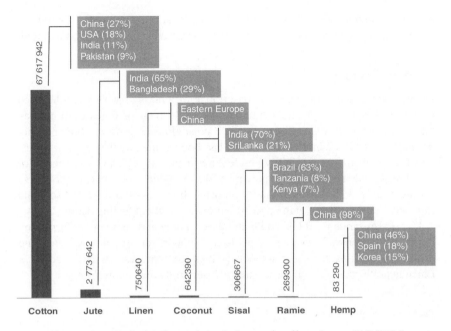

Figure 5.11 *Production (in metric tons) of some plant fibres. Source: FAO (2004)*

fibre is most often spun into thread and used to make textiles, which are the most widely used natural-fibre fabrics in clothing today.

The structure of cotton fibres is similar to that of the secondary walls (S1, S2 and S3) of a wood cell. Cotton is constituted of 98% cellulose with almost no lignin (no middle lamella gathering all the fibres together). The natural wax on the surface of the fibres is very useful for threading, as it lubricates their contact with the machines minimizing the fibre damage. Long fibres are used for textile applications. Shorter fibres (linters) of cotton are used as a cellulose source of high

Table 5.2   *Characteristics of the most common plant fibres*

		Linen	Ramie	Hemp	Jute	Sisal	Coconut	Cotton
Density	g·cm⁻³	1.54	1.56	1.07	1.44	1.45	1.15	1.5–1.6
Cellulose content	%	64–71	83	78	61–71	67–78	43	87–99
Microfibril angle	°	10	7.5	6.2	8	20	45	
Diameter	μm	5–76	16–126	10–51	25–200	7–47	12–24	10–20
Length	mm	4–77	40–250	5–55		0.8–8	0.3–1	10–55
Shape factor (L/D)		1700	3500	960	110	100	35	2000
Modulus of elasticity	GPa	12–85	60–130	35	25–30	9–21	4–6	5–13
Elongation at break	%	1–4	1.2–3.8	1.6	1.5–1.8	3–7	15–40	7–8
Strength at break	MPa	600–2000	400–1000	390	390–770	350–700	130–175	290–600

quality for biomaterials. It is used for the chemical transformation of cellulose (regeneration, esters and ethers, see Section 5.4) for which high purity is required.

Linen or Flax Fibre

As used today, the word 'linen' is descriptive of a class of woven textiles used in homes. Linens were manufactured almost exclusively of fibres from the flax plant *Linum usitatisimum*. Today flax is a prestigious, expensive fibre and only produced in small quantities. Flax fibres can be identified by their typical nodes, which account for the flexibility and texture of the fabric. The cross-section of the fibre is made up of irregular polygonal shapes, which contribute to the coarse texture of the fabric. When adequately prepared, linen has the ability to absorb and lose water rapidly. It can gain up to 20% moisture without feeling damp.

Linen uses range from bed and bath fabrics, home and commercial furnishing items (wallpaper/wall coverings, upholstery, support for oil paintings, etc.) and prestigious clothing, to industrial products (luggage, canvases, sewing thread, etc.). Linen is preferred to cotton for its strength, durability and archival integrity.

Jute

Jute is a long, soft, shiny vegetable fibre that can be spun into coarse, strong threads. It is produced from plants of the genus *Corchorus*. Jute is one of the cheapest natural fibres, and is second only to cotton in amount produced and variety of uses. It can be grown in 4 to 6 months. The jute hurd (inner woody core or parenchyma of the jute stem) is in the middle between textile fibre and wood. Therefore, it has good potential to fight against deforestation by industrialisation. Jute is the second most important vegetable fibre after cotton; not only for cultivation, but also for various uses. Jute is used chiefly to make cloth for wrapping bales of raw cotton, and to make sacks and coarse cloth. The fibres are also woven into curtains, chair coverings, carpets, rugs, and backing for linoleum.

While jute is being replaced by synthetic materials in many of these uses, some uses take advantage of its biodegradable nature, where synthetics would be unsuitable. Examples of such uses include containers for planting young trees which can be planted directly with the container without disturbing the roots, geotextiles, which are lightly woven fabrics made from natural fibres that is used for soil erosion control, seed protection, weed control and many other agricultural and landscaping uses. Geotextiles can be used for more than a year and biodegradable jute geotextile left to rot on the ground keeps the ground cool and is able to make the land more fertile.

Hemp

Hemp is the common name for *Cannabis sativa* cultivated for industrial (non-drug) use. Licenses for hemp cultivation are issued in the European Union and

Canada. Hemp grows quickly and produces strong fibres. In the past hemp was widely used for canvas (please note the same etymology) and other articles such as carpets and rope. The ultimate fibre is flatter, less regular and more lignified than linen. In Europe, the major application of hemp is in plastic–natural fibres composites as filler, mainly for the automotive industry. The microscopic protuberances appearing on the surface of the fibre represent an advantage for stronger mechanical anchorage with the plastic matrix. Another recent application is the fabrication of insulating mats for housing.

Ramie

Ramie (*Boehmeria nivea*) is one of the oldest fibre crops, principally used for fabric production, even mummy cloths, because of the non-fibrous material with antifungal and antibacterial properties. It is a bast fibre, and the part used is the bark (phloem) of the vegetative stalks. Unlike other bast crops, ramie requires chemical processing to de-gum the fibre (up to 25% mass loss).

Ramie is not as durable as other fibers, and so is usually used as a blend with other fibres such as cotton or wool. It is known especially for its ability to hold shape, reduce wrinkling and shrinking, and introduce a silky lustre to the fabric appearance. However it will not dye as well as cotton, but its white colour is useful in the textile industry. It is similar to flax in absorbency, density and microscopic appearance. Because of its high molecular crystallinity, ramie is strong, but stiff and brittle and will break if folded repeatedly in the same place; it lacks resiliency and is low in elasticity and elongation potential. When wet, it exhibits greater strength. Spinning the fibre is made difficult by its brittle quality and low elasticity; and weaving is complicated by the hairy surface of the yarn, resulting from lack of cohesion between the fibres. The greater utilization of ramie depends upon the development of improved processing methods. Ramie is currently used to make products such as industrial sewing thread, packing materials, fishing nets and filter cloths. Shorter fibres and waste are used in paper manufacture.

Sisal

Sisal is an agave (*Agave sisalana*) that yields a stiff fibre used in making rope. Sisals are sterile hybrids of uncertain origin; although shipped from the port of Sisal in Yucatan Mexico (thus the name), they do not actually grow in Yucatan, which cultivates, at present, henequen (*Agave fourcroydes*). Sisal plants consist of a rosette of sword-shaped leaves about 1.5 to 2 m tall. The sisal plant has a 7–10 year life span and typically produces 200–250 commercially usable leaves. Each leaf contains around 1000 packs of fibres. The fibre element, which accounts for only about 4% of the plant by weight, is extracted by a process known as decortication. Sisal is valued for cordage use because of its strength, durability, ability to stretch, affinity for certain dyestuffs, and resistance to deterioration in saltwater. Sisal is used by industry in three grades. The lower grade fibre is processed by the paper industry because of its high content of cellulose and hemicelluloses. The medium grade fibre is used in the cordage industry for making ropes. The higher-grade fibre, after treatment, is converted into yarns and used by the carpet industry.

Products made from sisal are being developed rapidly, such as furniture and wall tiles made of resonated sisal. A recent development has even expanded the range to car parts for cabin interiors. Other products developed from sisal fibre include spa products, cat scratching posts, lumbar support belts, rugs, slippers and cloths. In recent years sisal has been utilized as a strengthening agent to replace asbestos and glass fibre, as well as an environmentally friendly component in the automobile industry.

Coconut Fibres

The coconut palm (*Cocos nucifera*) is grown throughout the tropical world, for decoration as well as for its many culinary and non-culinary uses; virtually every part of the coconut palm has some human use. Coir (the fibre from the husk of the coconut) is used in ropes, rugs and mats, brushes and as stuffing fibre. It is also used extensively in horticulture for making potting compost.

5.3   Isolated and Modified Biopolymers as Biomaterials

Plants are wonderful chemical reactors that fabricate complex macromolecules. These compounds are located in the cell wall (e.g. cellulose, lignin, hemicelluloses and pectin) or they constitute the energy stocks (e.g. starch) and even they have specific functions (e.g. proteins). Most of these biopolymers are useful for making industrial biomaterials.

In order to develop valuable applications, extraction methods for these biopolymers are necessary. The isolation of starch is one of the most simple, as it requires essentially physical methods for extraction and purification, wet milling being the most common. On the other hand, the cell wall components need a chemical treatment to break the covalent bonds of the lignin–carbohydrate complex (LCC) to liberate the cellulose fibrils. This can be done either with sulphurous acid containing solutions of hydrogen sulphites, or with solutions of sodium hydroxide and sodium sulphate. Such hot treatments reduce wood chips to fibres by using a mild mechanical action. To further purify the raw pulp, a multistage refining procedure is needed in which alkali and oxidizing agents remove residual lignin. Finally, extractions with cold or hot alkali are used to remove pentanes and oligosaccharides. Pure cellulose (99%) can thus be obtained.

The characteristics of the isolated biopolymers depend on their structure. Cellulose and amylose are linear polymers, whereas amylopectin, pectin and hemicelluloses are branched polymers. Pectin and amylopectin contain carboxylic groups, which make interactions with water molecules very important. Amylose has a helix structure, whereas the cellulose molecule looks like a ribbon. The interactions with water and other neighbouring molecules are therefore different.

One of the most important aspects of these biopolymers is the fact that they cannot be used directly for thermoplastic applications. They are not reticulated polymers, but despite their linear or slightly branched structure, they suffer from two disadvantages. First, their structure contains relatively fragile bonds or fragile configurations that break with high temperatures. This is the case with the glycosidic bond or the C—O—C bond in the saccharide cycle of the polysaccharides. The tertiary structure of proteins is altered with heat as well. The second disadvantage is that they form numerous intermolecular hydrogen bonds resulting, in some cases, in very rigid structures. A lot of energy would be necessary to break these bonds and make the macromolecules flow. Polysaccharides decompose under the action of this energy before the polymer attains the molten state. This is particularly true in the case of cellulose and it is the reason why paper burns instead of melting when the temperature increases. In the case of starch, thermoplasticity only occurs with the help of an external plasticizer (water, glycerol, etc.), which disrupts the hydrogen bonding of the biopolymer.

In the following paragraphs, we will present the technologies needed to overcome these and other disadvantages. In this way, the natural polymer structure becomes profitable for industrial applications as biomaterials. We have selected only the main biopolymers currently used in commercial products.

5.3.1 Cellulose

The ribbon-like structure of the cellulose molecule (Figure 5.1) favours its organisation in oriented packs of about 50 to 100 molecules. If the organisation is completely regular, crystallites are formed. In nature it is found that the crystallinity rate approximates 50% in most species: wood, cotton, etc. The same molecule therefore shares both crystalline and amorphous regions. The fibrous structure of cellulose is maintained even after the chemical processes of pulping from wood. Indeed, pulping attacks and dissolves the lignin–carbohydrate complex, i.e. the middle lamella. The cell walls, which are constituted essentially of cellulose fibres, are thus isolated without severe chemical degradation. Such fibrous structure is useful for common bulk materials like paper or absorbing cushions (the so-called non-wovens). Pure celluloses are also used in high added value applications like hydrogels, stationary phases for chromatography or pharmaceutical formulations.

The potential of cellulose can be multiplied after chemical modification. The esterification or the etherification of its hydroxyl groups leads to new biopolymers with very different properties. They will be described later.

Let us first consider a particular case of the modification of the cellulose molecule: regeneration. Two types can be distinguished: with and without chemical modification.

Regeneration of Cellulose with Chemical Modification

In the *viscose* process, a reaction is carried out between cellulose and CS_2 in an alkaline solution to form a viscous solution of cellulose xanthate. The resulting

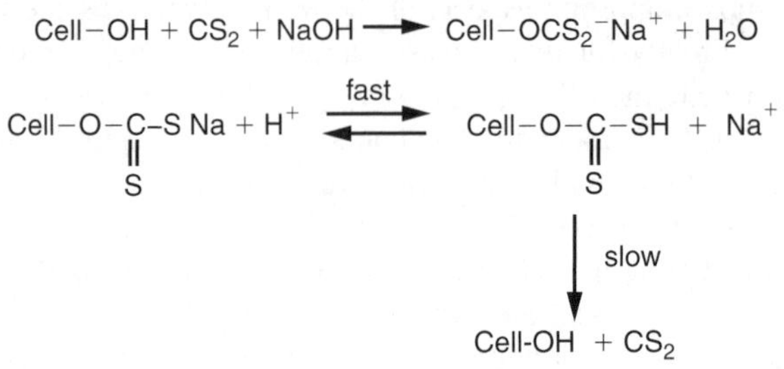

$$\text{Cell-OH} + CS_2 + NaOH \longrightarrow \text{Cell-OCS}_2{}^- Na^+ + H_2O$$

$$\text{Cell-O-}\underset{\underset{S}{\parallel}}{C}\text{-S Na} + H^+ \underset{\longleftarrow}{\overset{fast}{\longrightarrow}} \text{Cell-O-}\underset{\underset{S}{\parallel}}{C}\text{-SH} + Na^+$$

$$\downarrow slow$$

$$\text{Cell-OH} + CS_2$$

Figure 5.12   *Main reactions taking place in the viscose process for cellulose regeneration*

solution is filtered to get rid of particles. It is then extruded to form continuous sheets or threads or individual beads. When plunged into a water solution containing sulphuric acid and other additives, cellulose xanthate is hydrolysed and converted back into cellulose (Figure 5.12).

If the extrusion profile is a tiny cylinder, the obtained product is a continuous thread for textile applications. The industrial name of this product is Rayon®, or 'artificial silk'. A brilliant fabric with good dying properties is thus obtained.

The viscose process has some variants: depending on the quality of the cellulose and the composition of the regenerating bath, special high added value products can be obtained: so-called *modal-polynosic* fibres, or *modal-high wet modulus* fibres, for instance.

If the extrusion profile is a sheet, the obtained product is Cellophane®. Again, some additives are added to the hydrolysing bath to obtain the transparency and plastic aspect of the sheet. Even though the tensile strength of this product is high, the shearing resistance is low, which makes an excellent film for packaging and conditioning of biscuits or candies.

The viscose process has been abandoned progressively, but not totally, because of environmental concerns, as CS_2 is toxic and can easily cause explosions. It remains, however, historically important in the field of the chemistry of cellulose. Besides, it must be noted that the viscose process has set standards of variety, quality and cost that any new process must at least equal, or even surpass. If not, the safety and environmental restrictions may cause the total abandonment of the viscose process worldwide.

The *carbamate* process, which uses urea as a modifying agent, represents a greener alternative to the xanthate process (Ekman *et al.*, 1984). However, the industrial scale-up has not yet been successful.

Regeneration of Cellulose without Chemical Modification

The processes and products described in this section use solvents for cellulose to obtain solutions of the biopolymer able to be filtered and extruded. No derivatization of the cellulose hydroxyl groups occurs and therefore precipitation takes place instead of chemical regeneration.

- Among the most ancient solvents for cellulose we find cuprammonium hydroxide and cupriethylenediamine, but only the former was used industrially to give the CUPRO fibre.
- Later, other cellulose solvents were developed, like the N-methylmorpholine-N-oxide (NMMO) in water, systems of lithium chloride with an aprotic polar solvent, dimethyl acetamide, dimethyl formamide and dimethyl sulfoxide, but again, only the former has been successful and led to the LYOCELL fibre.
- More recently, ionic liquids capable of dissolving cellulose have been prepared. The best known is 1-N-butyl-3-methylimidazolium chloride (BMIMCl). No significant industrial development of yarn fabrication with this process is known so far.

The reasons why some solvents have not found industrial outlets include poor mechanical properties of the fibre (too rigid or low elasticity), difficulties in the control of all the process parameters and in the recycling of the solvent.

All the cellulose regeneration processes, with or without chemical modification, cause cellulose molecules to organise in a different crystalline form, called *cellulose II* and sometimes *cellulose IV* (especially in MODAL-HWM fibres)

It is worth noting that the mercerisation process, born in the 19th century, produces a *cellulose II* structure too, but without dissolution of the fibres and therefore with no reshaping. Cotton fibres are soaked in a concentrated (19%) NaOH solution then washed. Mercerised cotton shows a softer touch and more brilliance than natural cotton.

Microcrystalline Cellulose

Microcrystalline cellulose (MCC) is obtained by a controlled acid treatment intended to destroy the molecular bonding in the amorphous zones of cellulose. Usually HCl or H_2SO_4 are used at $110\,^{\circ}C$ for 15 min over native cellulose or regenerated cellulose. Colloidal gels are thus obtained showing thixotropy. MCC is used in the preparation of pharmaceutical compressed tablets due to its binding and disintegration properties.

In the ice cream industry, MCC is used to avoid the formation of big ice crystals, thus increasing unctuousness.

MCC is also used as a rheological modifier in water-based paint and toothpaste.

In chemistry laboratories, MCC is used as support for chromatography (column, thin layer, etc.).

5.3.2 Cellulose Esters

Cellulose esters are usually classified in organic and inorganic depending on the acid that is used to esterify the hydroxyl groups of cellulose.

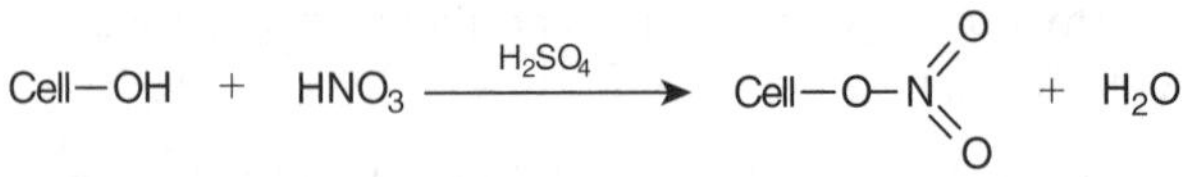

$$Cell-OH \;+\; HNO_3 \xrightarrow{\;H_2SO_4\;} Cell-O-N\!\!\overset{O}{\underset{O}{\diagdown}} \;+\; H_2O$$

Figure 5.13   *Synthesis of cellulose nitrate*

Cellulose Inorganic Esters

Cellulose nitrates are the most important inorganic esters of cellulose (Figure 5.13). Depending on the degree of substitution (DS), i.e. the average number of hydroxyl groups modified in a unitary glucose unit, the cellulose nitrates range from a resistant but inflammable polymer for film and photographic applications (celluloid grade, DS = 1.9), through a polymer for lacquers (DS = 2.05 − 2.35), to a powerful explosive that burns spontaneously in air (gun cotton, DS = 2.7).

The industrial fabrication of vegetable parchment usually sold as paper for baking passes through the formation of another group of cellulose inorganic esters, cellulose sulphates. Thus, a continuous sheet of paper is immersed for several seconds in a concentrated sulphuric (65–75%) acid bath maintained at low temperature. Sulphates are formed and acid hydrolysis of the cellulose fibres starts. The sheet passes through several rinsing baths to hydrolyse the sulphates and to eliminate the sulphuric acid (Figure 5.14). A continuous matrix of gellified cellulose is formed on both sides of the paper, which protects the inner mat of cellulose fibres. The latter ensures rigidity and mechanical resistance, whereas the continuous matrix shows extraordinary hydrophobic and lipophobic properties as well as high resistance to temperature. Kitchen parchment is then resistant to humidity and to greasy food and it is perfectly adapted for cooking.

Cellulose Organic Esters

The synthesis of cellulose organic esters can be accomplished in many ways. The acylation of the hydroxyl groups of cellulose require strong agents such as acid chlorides (Figure 5.15) or acid anhydrides. The former are preferred for long fatty

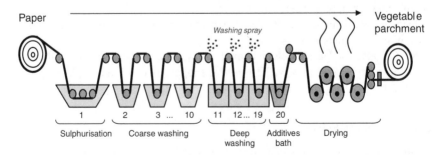

Figure 5.14   *Process for the production of vegetable parchment*

$$\text{Cell}{-}\text{OH} \ + \ \text{R}{-}\!\!\overset{\displaystyle O}{\underset{\displaystyle Cl}{\big\langle}} \ \longrightarrow \ \text{Cell}{-}\text{O}{-}\underset{\displaystyle O}{\overset{\displaystyle \|}{C}}{-}\text{R} \ + \ \text{HCl}$$

Figure 5.15   *Synthesis of cellulose organic esters by reaction with fatty acid chlorides*

chains, but it is necessary to use a strong base such as pyridine to neutralize the HCl formed, which can cause extensive degradation of the biopolymer. Other systems without pyridine have been proposed to limit the degradation of cellulose: use of partial vacuum (Kwatra *et al.*, 1992) and a dry nitrogen gas flow (Thiebaud and Borredon, 1995) to take HCl out of the reactor.

Nevertheless, the use of fatty acids (>C6) is still possible, but only with the use of a coreagent that forms new stronger entities *in situ*. Among them we can cite trifluoroacetic anhydride (Hamalainen *et al.*, 1957) and *N*,*N*-dicyclohexylcarbodiimide (Samaranayake and Glasser, 1993).

Cellulose acetates are by far the most important organic esters. The diacetate has DS = 2.4 and is fabricated either in filament form for fibres or in powder form to melt. Diacetate filaments are obtained by dissolution in acetone, extrusion through a spin and then evaporation of the solvent. The obtained fibres are used in textiles (called simply 'acetate') and in cigarette filters (tow). The triacetate (DS = 2.9) finds application in brilliant textiles easy to dye.

The industrial preparation of cellulose diacetate employs acetic anhydride with sulphuric acid as catalyst. The reaction is conducted at low temperature and cellulose starts to dissolve in the acetylation bath as the reaction progresses. The reaction is conducted until practically full acetylation. The homogeneous solution obtained is then hydrolysed to reduce the DS to 2.4. Precipitation in dilute acetic acid, then washing with water and finally drying produce cellulose acetate flakes.

The dramatic reduction of hydrogen bonding renders cellulose acetates thermoplastic. However, their softening points are very close to their decomposition temperatures (around 300 °C). The use of external plasticizers (e.g. phtalates) is often used to process cellulose diacetate in plastic applications (eyewear, screwdriver handles, etc.).

The cellulose mixed esters (acetate butyrate, acetate propionate) are used in the same plastic applications. The irregularity introduced by two different substituents diminishes the glass transition to around 120 °C. Moreover, the combination of a short side chain such as acetate and a long one coming from a fatty acid (in a 2.4:1 ratio respectively) results in very interesting biomaterials since they are as hydrophobic as simple fatty esters, but with better mechanical properties (Figure 5.16) and higher glass transition (Figure 5.17). Their biodegradability is also increased due to changes in crystallinity.

Cellulose esters of long-chain carboxylic acids (up to C_{20}) are described as interesting thermoplastic materials. An exhaustive review of all the synthesis methods

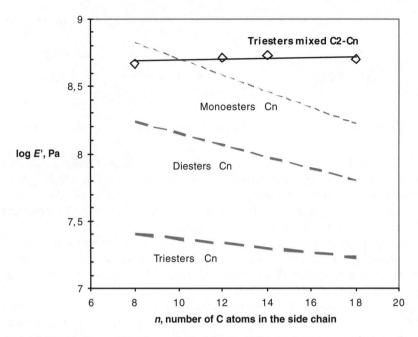

Figure 5.16 *Modulus of elasticity (E') of simple fatty esters and mixed acetic-fatty triesters of cellulose (DMA measurements)*

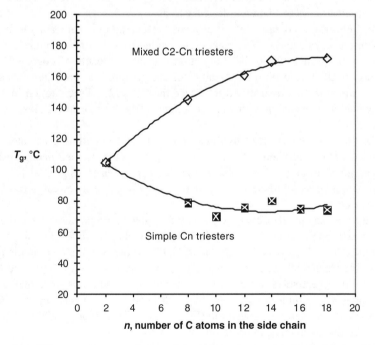

Figure 5.17 *Glass transition temperatures of simple fatty triesters and mixed acetic-fatty triesters of cellulose*

has recently been done (El Seoud and Heinze 2005). These derivatives present softening points between 70 and 250 °C depending on the DS and the number of carbon atoms in the acyl chain (Vaca-Garcia *et al.*, 2003). The higher the DS value the lower the softening point. Nevertheless, the difficulty in obtaining high DS values without using toxic chemicals is a limiting factor for the development of cellulose long chain esters in the plastics industry.

Nevertheless, cellulose fatty esters with low DS values show other qualities, such as a high hydrophobicity. The development of water-repellent cellulosic materials (i.e., cotton, wood), has led to interesting applications in the textile and wood industries. For instance, the direct esterification of timber with fatty acids (and their derivatives) has resulted in extraordinary outdoor durability and resistance to biological attack (e.g. rotting, termites). Industrial exploitation of this technology has recently been conducted in France [WoodProtect® by Lapeyre (Magne *et al.*, 2003)]. In this case, the water-repellence conferred to wood and the lack of recognition from predator enzymes account for these properties.

In general, after esterification with organic acids there is a change in many of the properties of the cellulosic fibres. These are collected in Table 5.3.

5.3.3 Cellulose Ethers

Cellulose ethers are the most produced cellulose derivatives. According to the type of substituent that replaces the hydroxyl function, the cellulose ether can be soluble in water or in organic solvents giving viscous solutions. When applied to water at about 1 to 2%, viscosity can rise up to 50 Pa·s (50 000 times the viscosity of pure water). They are therefore used as rheological modifiers in numerous formulations and are good substitutes for xanthan gum. The raw material can be either wood pulp or cotton linters. The latter, which have a higher degree of polymerisation, are used for the production of cellulose ethers with high viscosity.

Carboxymethyl Cellulose

Carboxymethyl cellulose (CMC) is a white solid, without odour and innocuous. Its sodium salt is more common, obtained by reaction of alkali cellulose with sodium chloroacetate instead of chloroacetic acid. The rheological modification properties of sodium CMC depend on the degree of substitution and on the degree of polymerisation. Commercial sodium CMC has DS values between 0.6 and 0.9, and is water soluble. Sodium CMC is widely used in many applications, not as biomaterial itself, but as a component of many materials. These include: additive in textiles (ironing, printing, anti-shrinking), paper (retention of fillers, coatings, glues for wallpaper), cement (retarder agents), etc.

Methyl Cellulose

Methyl cellulose (MC) is a yellowish or white solid, with no odour or flavour. It is obtained by reaction of alkali cellulose with methylene chloride. It is soluble in

Table 5.3 Main changes observed in cellulosic fibres after esterification of the hydroxyl groups with organic acids

	Cellulose fibres	Esterified cellulose fibres	References
Thermoplasticity	Do not melt. Thermal degradation above 200 °C	Show apparent melting point	Shiraishi et al., 1983
Hydrophobicity	Hydrophilic. Swell in water	Water resistance. Dimensional stability	Rowell and Keany, 1991
Biological resistance	Easily attacked by insects and fungi	Resistance to termites and fungi	Rowell, 1997
Biodegradability	Readily biodegradable	Retarded biodegradation	Glasser et al., 1994
Solubility	Insoluble in ordinary organic solvents	Soluble in various organic solvents	
Inflammability	Burn easily	Do not burn readily	Rowell, 1997
Weathering resistance	Fast degradation due to water absorption (rotting and biological attack)	Increased UV and water resistance	Rowell, 1997

water with DS-values between 1.4 and 2. MC and its derivatives (mixed ethers with hydroxyethyl, hydroxypropyl and hydroxybutyl) form gels when the water solutions are heated. The temperature of gelling depends on the DS, concentration, degree of polymerisation and the presence of salts or organic solvents. MC appears in the formulations of different materials, similar to those of CMC.

Ethyl Cellulose

Ethyl cellulose (EC) is a water-insoluble cellulose ether. It is produced by reaction of alkali cellulose with ethylene chloride. It has film-forming and thermoplastic properties. As a plastic, it can be processed by extrusion and injections. It is hard, stiff and with good resistance to impact. It is soluble in the molten state with other thermoplastics. As for its film-forming properties, it is used in the formulation of varnishes, inks and glues. It forms removable coatings.

5.3.4   Starch

Starch is the main energy reserve of superior vegetal plants. It is found in big quantities in wheat, potato, corn and manioc. Starch is a homopolymer (99%) of D-anhydroglucopyranose units. Nevertheless, two different configurations exist: amylose and amylopectin (Figure 5.18). The proportion of branched polymer is 70–80% (Park *et al.*, 2007). Native starch is present in the form of partially crystalline (25–40%) granules (up to 100 μm diameter), showing a complex structure, which has been the object of thousands of scientific papers.

The starch grains are insoluble in water at room temperature. At 50–60 °C, starch absorbs water reversibly and hydrogen bonding is reduced. Above 60 °C (this temperature depends on the native source) the structure of the starch is modified irreversibly, crystallinity disappears and gelatinisation occurs.

The paper industry is the main non-food outlet for starch and consumes 17% of the European starch production. Starch–cellulose–starch bonds are created and contribute to the internal cohesion of the paper sheet.

Thermoplastic materials may be obtained by extrusion in the presence of water at temperatures as high as 160–200 °C. If an external plasticizer (i.e., glycerol or sorbitol) is added, the glass transition temperature decreases. Even though these materials are fully biodegradable, their affinity with water is a big inconvenience when they are considered for the replacement of traditional plastic materials. Blends and chemical modifications are required to overcome these problems.

Finally, starch is the main source of glucose for fermentation processes intended to obtain biosynthesized polymers, which will be described later.

Starch Derivatives

The *oxidation* of starch leads to the formation of carbonyl and carboxyl groups in the polymer chain. Depolymerisation also occurs and starch turns yellowish.

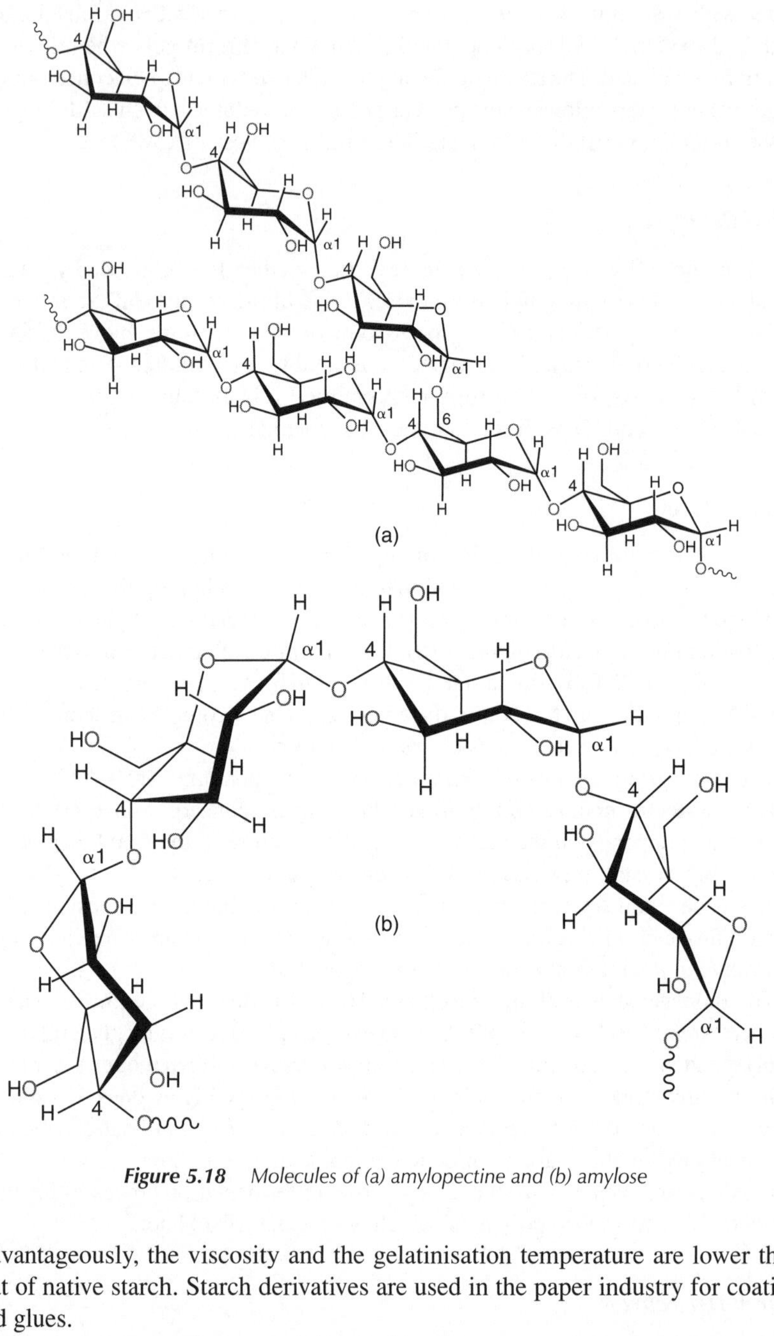

Figure 5.18 *Molecules of (a) amylopectine and (b) amylose*

Advantageously, the viscosity and the gelatinisation temperature are lower than that of native starch. Starch derivatives are used in the paper industry for coating and glues.

Non-ionic starch ethers are used in the food industry to avoid water release from frozen food and in the paper industry as a coating agent.

Cationic starch ethers are used in the paper industry to increase the cohesion and rigidity of cellulose fibres and as a flocculant for the selective separation of negatively charged particles.

Reticulated starch, obtained by reaction with bifunctional reagents, increases the water retention capacity and diminishes the swelling of starch grains, thus increasing the mechanical and thermal resistance. This is particularly interesting for the fabrication of highly absorbent nappies.

Starch esters are thermoplastic materials with properties that are somewhat similar to that of cellulose esters. In particular, starch acetate (DS < 0.2) is used as a coating agent for paper and as a food or detergent additive.

5.3.5   Chitin and Chitosan

Chitin is a polysaccharide constituted of *N*-acetylglucosamine, which forms a hard, semitransparent biomaterial found throughout the natural world. Chitin is the main component of the exoskeletons of crabs, lobsters and shrimps. Chitin is also found also in insects (e.g. ants, beetles and butterflies), and cephalopods (e.g. squids and octopuses) and even in fungi. Nevertheless, the industrial source of chitin is mainly crustaceans.

Because of its similarity to the cellulose structure, chitin may be described as cellulose with one hydroxyl group on each monomer replaced by an acetylamine group. This allows for increased hydrogen bonding between adjacent polymers, giving the polymer increased strength.

Chitin's properties as a tough and strong material make it favourable as surgical thread. Additionally, its biodegradibility means it wears away with time as the wound heals. Moreover, chitin has some unusual properties that accelerate healing of wounds in humans. Chitin has even been used as a stand-alone wound-healing agent.

Industrial separation membranes and ion-exchange resins can be made from chitin, especially for water purification. Chitin is also used industrially as an additive to thicken and stabilize foods and pharmaceuticals. Since it can be shaped into fibres, the textile industry has used chitin, especially for socks, as it is claimed that chitin fabrics are naturally antibacterial and antiodour (www.solstitch.net). Chitin also acts as a binder in dyes, fabrics and adhesives. Some processes to size and strengthen paper employ chitin.

Chitosan is produced commercially by deacetylation of chitin. It is a linear polysaccharide composed of randomly distributed β-(1-4)-linked D-glucosamine (deacetylated unit) and *N*-acetyl-D-glucosamine (acetylated unit). The degree of deacetylation in commercial chitosans is in the range 60–100% (Figure 5.19).

The amino group in chitosan has a pKa value of about 6.5. Therefore, chitosan is positively charged and soluble in acidic to neutral solutions with a charge density dependent on pH and the deacetylation extent. In other words, chitosan readily binds to negatively charged surfaces such as mucosal membranes. Chitosan enhances the transport of polar drugs across epithelial surfaces, and is biocompatible

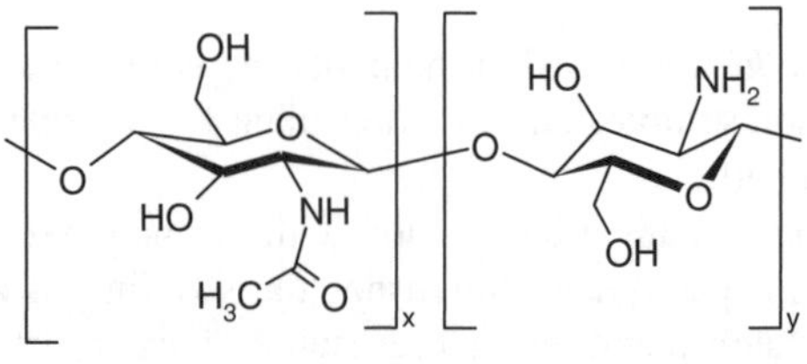

Figure 5.19   *General formula for chitosan and chitin. In the case of chitin* y = 0

and biodegradable. Purified qualities of chitosan are available for biomedical applications.

Chitosan possesses flocculating properties, which are used in water processing engineering as a part of the filtration process. It may remove phosphorus, heavy minerals and oils from the water. In the same manner, it is used to clarify wine (as a substitute for egg albumin) and beer.

5.3.6   Zein

Zein is a vegetable protein obtained from corn gluten meal. Pure zein powder is odourless, tasteless, hard and water-insoluble. Since it is edible, it finds applications in processed foods and pharmaceuticals, in competition with chitosan and chitin.

Historically it has been used in the manufacture of a wide variety of commercial products, including coatings for paper cups, soda bottle cap linings, clothing fabric, buttons, adhesives and binders. It is now used for the encapsulation of foods and drugs.

The main barrier to greater commercial success has been its historic high cost until recently. Some believe the solution is to extract zein as a by-product in the manufacturing process for bioethanol. From a high-value added perspective, Chinese researches have conducted intensive research on the use of zein as a biomedical material (Dong *et al.*, 2004).

5.3.7   Lignin Derivatives

A portfolio of biomaterials has been obtained from lignin derivatives. First, a lignophenol derivative is obtained, which contains a diphenylpropane unit formed by binding a carbon atom at the *ortho* position of a phenol derivative to a carbon atom at the benzyl position of a phenylpropane fundamental unit of lignin, and binding an oxygen atom of the hydroxyl group and a β-positional carbon atom under alkali conditions to obtain an arylcoumaran derivative (Funaoka, 2005). The latter can be reticulated to form polymers shaped under hot-moulding.

A similar strategy consists in the liquefaction of biomass with phenol under acidic conditions to obtain phenolic monomers. The type of monomers obtained vary greatly (Lin *et al.*, 2001). These monomers can be reticulated with the help of temperature and formaldehyde to obtain novolac- or resol-type resins.

To our knowledge, none of these biomaterials has shown significant biodegradability.

5.4 Agromaterials, Blends and Composites

5.4.1 Agromaterials

The food processing industry produces a large amount of waste and coproducts rich in fibres. Their nature and the type of biopolymers contained in them vary greatly. The three main food industries: vegetable oil, starch and sugar, produce million of tons of oilcake, stalks, pulp and bagasse (Table 5.4). Most of these residues are used for animal feed. However, their high cellulose content and their low price make them a source of choice for the fabrication of materials that have, for two decades, been called 'agromaterials'.

The availability of these residues depends on regional agricultural productions and on the commitment of agricultural cooperatives in favour of agromaterials. For instance, in France, at present, there are corn crops dedicated exclusively to the fabrication of thermoplastic agromaterials.

In this case the biopolymers are directly plasticized by thermomechanical means, and transformed through the classical forming technologies of the plastic industry: injection-moulding, extrusion and thermoforming. These agromaterials keep a natural aspect, they are sensitive to the atmospheric conditions like wood, but they have no shape restrictions (Figure 5.20).

Agromaterials demonstrate that it is possible to profitably transform raw agricultural products without separation and that almost all non-cellulosic biopolymers can be plasticized *in situ* to constitute a natural continuous matrix for cellulosic fibres.

Destructuring the native organization of the raw agricultural products is possible with a combination of thermal, mechanical and chemical effects in an extruder (Figure 5.21). In the first constraint zone, the matter is roughly crushed. In the second, a compression (up to 20 bars) in the reverse screw induces a non-degradative break

Table 5.4 *Average composition (%, dry matter basis) of the three common feedstocks used for the production of agromaterials*

	Sunflower Oilcake	Sugar Beet Pulp	Whole Corn Plant
Lipids	1	0	3
Sugars	6	0	7
Ash	7	5	5
Cellulose	12	23	21
Hemicelluloses & pectins	17	48	24
Lignin & phenolics	13	2	3
Starch	1	0	30
Protein	34	7	7

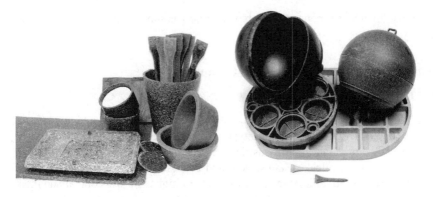

Figure 5.20   *Examples of thermoplastic agromaterials obtained from sunflower oilcake and whole corn plant*

of the structure when the moisture content and the temperature are between 20 and 30% and 110 °C and 130 °C, respectively. Water plays the roles of plasticizer and lubricant and avoids degradation under high shear. The same phenomenon occurs with a higher compression and a lower shear in the die to complete the transformation.

The starch plasticization is obtained by gelatinization of the grains in low moisture conditions leading to the 'melting' of the starch grains. This is the key phenomenon in the transformation of the whole corn plant. The plasticized starch forms a continuous matrix in which the defibrated fibers are embedded, as showed schematically in Figure 5.22.

The transformation of sunflower oilcake is similar. The globulin corpuscles are denatured in low moisture conditions to form a continuous matrix. The real difference lies in the structure of the peptide chains compared to starch. A lot of different interactions take place between the proteins and their texturation does not result in a simple 'thermoplastic' flow, as happens for starch.

Sugar beet pulp is made of primary cell walls. The breakage of the structure has to be done at a smaller scale and a higher energy is needed to break some

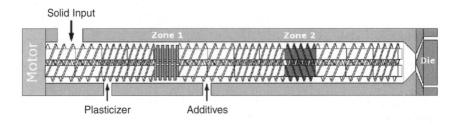

Figure 5.21   *Twin-screw extruder configuration for the transformation of raw agricultural products*

Starch granules 'melting' and
ligno-cellulosic fibres defibration

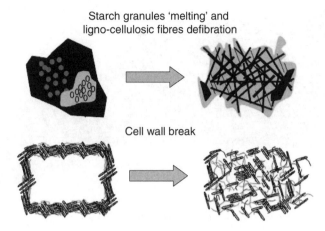

Cell wall break

Figure 5.22 *Schematic representation of the transformations taking place in the twin-screw extruder*

of the covalent bonds between the cell-wall polysaccharides. The use of the die is therefore necessary to increase the residency time in the extruder. The final structure consists of cellulose microfibers embedded in a pectin and hemicellulose matrix.

Injection moulding is the first targeted forming technology for agromaterials. Without any important technological update, injection-moulded objects can be obtained quickly and easily. At the end of the twin-screw extrusion process, the obtained granulates are stable and mouldable by injection. Typically, the shaped materials are dense (density $\cong$ 1.4), relatively stiff and brittle (flexural modulus $\cong$ 4 GPa; strength at break $\cong$ 17 MPa) and water sensitive. The latter is a profitable characteristic for the biodegradation of the agromaterials, which reaches 100% in a very short time. A particular interest is in the pricking out of plants. A pot made of any of these agromaterials is strong enough to ensure mechanical support for growing small plants. When they are big enough, the plants can be transplanted to the soil without removing the pots. The latter will biodegrade and provide nutriments to the plant, which has been demonstrated to grow faster and bigger.

In the particular case of protein-rich granulates (oilcakes), thermomoulding is often more suitable because it takes advantage of their cross-linking abilities (Rouilly *et al.*, 2005). The mechanical properties of these materials are lower, but the advantage is that oilcake can be used as it is, i.e., at its equilibrium moisture content, and the materials are more resistant to water.

Regarding its microstructure, sugar beet pulp is the best candidate to extrude films (thickness > 200 μm) (Rouilly *et al.*, 2006). In this case, glycerol is used as an external plasticizer (Figure 5.23).

Figure 5.23   *Film obtained by extrusion of sugar beet pulp*

5.4.2   Blends of Synthetic Polymers and Starch

The blends of starch and a synthetic polymer (usually polyethylene) are products of commercial importance. Two families of blends are obtained: those using dehydrated starch pellets and those using gelatinised or thermoplastic starch. In both cases, the mixture with the synthetic polymer is done by extrusion. Further processing by moulding or blowing is still possible, depending on the kind of starch used.

Dehydrated starch acts as a filler and diminishes the mechanics properties of the polymer. It is not possible to add more than 20%. Normally, in practice, the rate oscillates between 6 and 10%. In order to get a better compatibility between starch and the polymer, starch might be rendered hydrophobic by chemical treatment, as described above for cellulose.

One of the principal objectives in these systems is biodegradability. In this regard, auto-oxidation agents and organometallic catalysts are added to the blend. They are intended to break the synthetic polymer chains up to the point where they became available to microorganisms, thus making starch accessible. Degradation initially starts at a low rate and is accelerated when exposed to degradation conditions (compost, etc.). The useful life of these blends is not intended to be too long anyway. In most cases, the material is biofragmented, but not fully biodegraded or bioassimilated when exposed to degradation. In order to reduce water absorption of starch blends, and therefore their mechanical properties, the addition of zein seems to be efficient (Gaspar *et al.*, 2005).

There are some examples at a commercial level of starch-containing blends:

- *Ecopolym* (Polychim). In this case the synthetic polymer is polyethylene and starch is present at 10%. This is associated with one catalyst that promotes the decomposition by oxidation and by cleavage of chains thanks to the radicals produced.

• *Ecostar* (St. Lawrence Starch Company). This product associates PE with a mixture of starch and auto-oxidant unsaturated fatty acids. The global content of starch is between 6 and 15%. The degradation process then follows two mechanisms: in the first, the starch is fragmented, then assimilated by microorganisms, whereas in the second, the interaction between the auto-oxidants and the metallic complexes from soil or water gives peroxides that attack the synthetic polymer chains.

These kinds of products are normally used in mulch films, bags and packing.

Other systems have been investigated. The combination starch/ polyester has been claimed to be fully biodegradable (Tokiwa *et al.*, 1993). Others are partially biodegradable, like starch/polyethylene/poly-ε-caprolactone blends (Corp, 2001) and their derivatives or combinations of starch and modified polyesters.

A particular mention goes to *Mater-Bi*, produced by Novamont, who have revolutionised starch-based biomaterials for two decades. The commercial success of this biodegradable and biocompostable plastic relies on two main factors: the scale economy that allows the reduction of costs, and the diversity of formulations to develop different end products (plastic bags, tableware, toys, etc.). More than 210 references in *Chemical Abstracts* are available on this (registered) keyword, and the number of patents related to different formulations and developments is also impressive. Mater-Bi can be essentially described as a blend of starch with a small amount of other biodegradable polymers and additives. The actual compositions are still known only by a very few people.

5.4.3 Wood Plastic Composites (WPC)

Vegetable fibres (including wood fibres) represent a good replacement solution for glass and carbon fibres for the reinforcement of composites based on a thermoplastic matrix. The advantages of vegetable fibres are economically and ecologically important:

• They are inexpensive and slightly abrasive for transformation tools.
• Their density is low and lightens the whole composite material.
• They have a limited environmental impact.

In 75% of the cases, wood fibres are the preferred filler for thermoplastic matrices. Their high availability may account for this. The resulting composite material is internationally known as WPC, which stands for wood plastic composite. They have been produced industrially since 1980 and the market has gradually increased in the last 10 years, especially in the United States, reaching 700 000 metric tons with an 11% increase rate per annum. In Europe, the WPC market is considered to be emerging. At the best estimations, it reached only 100 000 metric tons in 2005 (Anonymous, 2006) and in Japan the market is even smaller. In America, more than 50% of the WPC is used for parquets and decking made with polyethylene. In Europe, it is the automobile sector that is preponderant with polypropylene-based composites.

For these two types of WPC, the most studied parameters are:

- The fibre/matrix adhesion
- The filler content
- The granulometry of the fibres
- The extrusion parameters
- The durability of the WPC when exposed to water, sunlight, fungi and insects.

The adhesion between the fibre and the matrix is by far the most crucial parameter of composite materials. The mechanical properties of WPC depend greatly on it and on the compatibility of the filler and the matrix. The only practical way to improve the adhesion is to add a coupling agent to the formulation, i.e., a molecule capable of establishing bonding between the filler and the matrix. Three kinds of bonding are involved: covalent, hydrogen and non-polar interactions. They all contribute to reduce the natural incompatibility between the hydrophilic fibres and the hydrophobic matrix. The most used coupling agent for polyolefine-based composites is MAPP, maleinated polypropylene. It is prepared from PP and maleic anhydride (Figure 5.24). The anhydride function reacts with wood fibres and the attached PP moiety is fully compatible with free PP.

The addition of MAPP at around 1–2% ensures perfect coverage of the wood fibres by polypropylene or polyethylene. The micrographs in Figure 5.25 clearly show the lack of adhesion between the fibres (in dark colour) and the PP matrix (in light colour). The mechanical properties are increased by at least 30% when a coupling agent is used.

For PVC composites, used in housing and in the automobile industry, no coupling agent is used because PVC is rather polar and compatible with wood fibres. Some research has been done and the positive effect of aminosilanes (Kokta *et al.*, 1990a) and poly(methylene(phenyl isocyanate)) (Kokta *et al.*, 1990b) has been demonstrated.

It must be pointed out that most WPC is not biodegradable material. The synthetic matrix remains a fossil resource with negligible biodegradation even if it is accompanied by natural fibres. Nevertheless, these materials show some

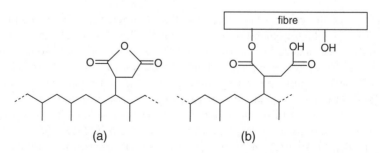

(a)   (b)

Figure 5.24   *(a) Maleinated polypropylene and (b) the adduct formed with vegetable fibres by covalent bonding*

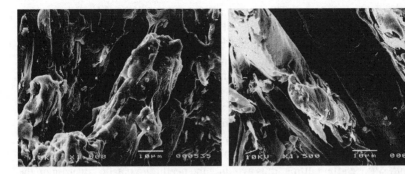

Figure 5.25 *Scanning electron micrographs of fracture surfaces of polypropylene/wood fibres composites. Left: with maleated PP; right: without MAPP*

advantages as well. Indeed, polypropylene needs to be reinforced with a filler when recycled in order to maintain its mechanical properties. The WPC is therefore a perfect solution to prolong of the use life of olefins. As for the recycling of WPC itself, only PVC/wood composites have demonstrated no loss of properties when thermomechanical recycling is applied (Augier *et al.*, 2007).

5.4.4 Wood-Based Boards

When wood sheets or particles of wood are bound together to get a more or less stiff board, lots of applications are derived. They depend essentially on the mechanical properties and on the density of the board. Secondarily, these applications also depend on the granulometry of the wood employed.

In a first classification, we can distinguish: (i) boards made with an external synthetic binder, such as urea-formaldehyde or phenol-formaldehyde resins for thermosets, and (ii) boards with internal natural binders. The products cited below do not constitute an exhaustive list.

Wood Boards with Synthetic Binder

Different commercial products are classified according to the size of the form of the raw matter (wood). All of them require the binder (around 10%) to be sprayed on the wood mat before undergoing hot-pressing (150–200 °C, 20–25 bar).

Wood sheets, which are obtained by 'unrolling' steamed wood rods with a blade, can be bound together to form thicker boards. The main characteristic of these 'sandwiches' is the alternation of the fibre orientation by 90° in every layer. The anisotropy of the whole board is thus limited in the length and in the width. Three, five or seven layers are usually used in these panels, called *plywood*. Often, the inner layers of plywood are replaced by OSB (see below) to reduce costs. The two outer layers of wood sheets ensure the aesthetic appearance of the panel.

Oriented strand board (OSB) is obtained from wood chips. The orientation is designed to simulate the characteristics of a wood panel, with limited swelling and higher resistance in the fibre direction. Chips are orientated by air with a blowing machine. These panels are commonly used on building sites, and they often make part of the walls in houses, especially in the United States. If no orientation is given, the panel is called *waferboard*.

When sawdust is used, particleboards are obtained. The higher granulometries are used for *fibreboard*, which is the material most used in the fabrication of affordable furniture and kitchens. The smallest particle sizes, i.e., wood flour, are used for *medium density fibreboard* (MDF), which has the advantage of showing a smooth surface after cutting, thus avoiding the need for a plastic covering on the edge (as particleboard does).

Urea-formaldehyde resins are used for applications in which the panel is not in contact with water. For applications with high level of humidity, the phenol-formaldehyde resins are required. In all cases, free formaldehyde constitutes a dangerous pollutant that is slowly released and can be particularly toxic in con-fined rooms. Recent research has been done to substitute this type of resin by natural binders, such as cross-linkable proteins (Silvestre *et al.*, 2000; Yang *et al.*, 2006).

Wood Boards without External Binder

In this case, water is sprayed on sawdust instead of a synthetic binder. Hot-pressing of the mat is carried out to partially hydrolyse some macromolecules contained in the wood as well as thermal degradation of free sugars and other small molecules. The degradation products contribute to the cohesion of the particles.

Depending on the pressure applied, more or less rigid boards are obtained. The ones with lowest density (0.15 to 0.5), processed with the lowest pressure, are called *insulation board*. As its name indicates, they are used in housing for thermal and phonic insulation of the walls and ceilings. In practice, they are associated with a more rigid material to obtain a less fragile product.

If the pressure used in the fabrication is increased, the density of the product is higher (0.5 to 1.45). *Hardboard* is thus obtained. This kind of board is used in furniture parts in which the mechanical resistance required is not very high, for instance, the back panel of inexpensive bookcases.

5.5   Biodegradable Plastics

5.5.1   Polyglycolic Acid (PGA)

Polyglycolide or olyglycolic acid (PGA) is a biodegradable, thermoplastic polyester. It can be prepared by polycondensation of glycolic acid or by ring-opening polymerisation of glycolide. Glycolic acid (or hydroxyacetic acid) is the smallest α-hydroxy acid (AHA) and is commonly isolated from sugar crops (sugarcane, sugar beets). Glycolide is the cyclic diester of glycolic acid. Synthesis of PGA from glycolide leads to higher molecular weights.

PGA has been known as a polymer capable of forming tough fibres; however, due to its hydrolytic instability its use has initially been limited. Currently, PGA and its copolymers with either lactic acid or with ε-caprolactone are widely used as a material for the fabrication of absorbable sutures.

Polyglycolide has a glass transition temperature between 35 and 40 °C and its melting point is around 225–230 °C, depending on the molecular weight. PGA also exhibits an elevated degree of crystallinity, around 45–55%, thus resulting in insolublity in water and in almost all common organic solvents. Only highly fluorinated solvents, like hexafluoroisopropanol and hexafluoroacetone sesquihydrate, dissolve PGA and are used to prepare solutions for melt spinning and film preparation. Fibres of PGA exhibit high strength and modulus and are particularly stiff.

Polyglycolide is characterised by hydrolytic instability owing to the presence of the ester linkage in its backbone. This is common to all the polyesters cited in this section. When exposed to physiological conditions, PGA is degraded by random hydrolysis and apparently it is also broken down by esterases. The degradation product, glycolic acid, is non-toxic and it can enter the tricarboxylic acid cycle, after which it is excreted as water and carbon dioxide. A part of the glycolic acid is also excreted in urine.

Studies using polyglycolide-made sutures have shown that the material loses half of its strength after two weeks and 100% after four weeks. The polymer is completely reabsorbed by the organism within four to six months.

The traditional role of PGA as a biodegradable suture material has led to its evaluation in other biomedical fields, such as tissue engineering or controlled drug delivery and even dental fillings.

Copolymers with lactic acid have been prepared to increase the biodegradability for medical applications.

5.5.2 Polylactic Acid (PLA)

Polylactide or polylactic acid (PLA) was discovered around 1900, but it has only now found a universal route to market in the form of biodegradable thermoplastic packaging. Commercial products like yoghurt pots, tableware and containers for liquids are available in commercial form. In ideal conditions, PLA can be mineralized in 60 days. Corn or wheat starches (or saccharose in some cases) are the common feedstock. Bacterial fermentation is used to produce lactic acid, which is dimerized to make the monomer for ring-opening polymerization. It can be easily produced in a high molecular weight form through ring-opening polymerization using most commonly a stannous octoate catalyst or tin(II) chloride.

Due to the chiral nature of lactic acid, several distinct forms of polylactide exist: poly-L-lactide (PLLA) is the product resulting from polymerization of L,L-lactide (also known as L-lactide). PLLA has a crystallinity around 37%, a glass transition temperature between 50 and 80 °C and a melting temperature between 173 and 178 °C. The polymerization of a racemic mixture of L- and D-lactides leads to the synthesis of poly-DL-lactide (PDLLA), which is not crystalline, but amorphous.

Polylactic acid can be processed, like most thermoplastics, into fibres (for example using conventional melt spinning processes) and film. The melting temperature can be increased by 40–50 °C by physically blending the polymer with PDLA (poly-D-lactide). The maximum effect in temperature stability is achieved when a 50:50 blend is used, but even at lower concentrations of 3–10% of PDLA, a substantial effect is achieved. In the latter case PDLA is used as a nucleating agent, thereby increasing the crystallization rate and the transparency. However, due to the higher crystallinity of this stereo-complex, the biodegradability decreases.

The physical blend of PDLA and PLLA can be used in other applications, such as woven shirts with better ironability, microwavable trays, hot-fill applications and even engineering plastics (blends with rubber-like polymers such as ABS). PLA is also currently used, like PGA, in a number of biomedical applications, such as sutures, dialysis media, drug delivery devices and tissue engineering.

NatureWorks LLC, a wholly owned subsidiary of Cargill Corporation, is the primary producer of PLA in the United States. Other companies involved in PLA manufacturing are Toyota (Japan), Hycail (The Netherlands), Galactic (Belgium) and several Chinese manufacturers. The price of PLA has been falling as more production comes online.

5.5.3 Polycaprolactone (PCL)

This is not a biomaterial as defined in the introduction to this chapter (not obtained from renewable resources), but due to its biodegradability it is often presented along with other biomaterials. Polycaprolactone (PCL) is a biodegradable polyester with a low melting point of around 60 °C and a glass transition temperature of about −60 °C. PCL can be prepared by ring opening polymerization of ε-caprolactone using a catalyst such as stannous octoate. It is also quite common to copolymerize ε-caprolactone with DL-lactide to reduce crystallinity.

Since PCL is compatible with a range of other materials, it can be mixed with starch to lower its cost and increase biodegradability or it can be added as a polymeric plasticizer to PVC.

The degradation of PCL *in vivo* is lower than that of PGA and PLA. Therefore, its use in long-term implantable devices has been studied.

The Solvay company produces PCL under the name of CAPA, and Union Carbide under the name of TONE.

5.5.4 Polyhydroxyalkanoates (PHA)

Polyhydroxyalkanoates (PHAs) are polyesters that were first isolated and characterized in 1925 by French microbiologist Maurice Lemoigne. They are produced by microorganisms (e.g. *Alcaligenes eutrophus* and *Bacillus megaterium*) in response to conditions of physiological stress. The polymer is primarily a product of carbon assimilation (from glucose, coming from starch) and is employed by microorganisms as a form of energy storage molecule to be metabolized when other common energy sources are not available. Poly-3-hydroxybutyrate (P3HB) is probably the most common type of PHA. Its microbial biosynthesis starts with the condensation of two molecules of acetyl-CoA to give acetoacetyl-CoA, which is subsequently reduced to hydroxybutyryl-CoA. This latter compound is then used as a monomer to polymerize PHB.

Many other polymers of this class are produced by a variety of organisms: these include poly-4-hydroxybutyrate (P4HB), polyhydroxyvalerate (PHV), polyhydroxyhexanoate (PHH), polyhydroxyoctanoate (PHO) and their copolymers.

Chemical structures of P3HB, PHV and their copolymer Polyhydroxybutyratevalerate (PHBV) have attracted much commercial interest as plastic materials because their physical properties are remarkably similar to those of polypropylene (PP), even though the two polymers have quite different chemical structures. While PHB appears stiff and brittle, it also exhibits a high degree of crystallinity, a high melting point of about 180 °C, but, most importantly, PHB is rapidly biodegradable, unlike PP. For this reason, PHB is being evaluated as a material for tissue engineering scaffolds and for controlled drug-release carriers.

Two major factors inhibiting widespread use of PHB lie in its production costs, which are one of the highest among the biopolymers, and its brittleness, since PHB as it is currently produced cannot handle high impact. There are also some concerns about how large quantities of PHB would affect the environment, as its consumption of fossil energy counterbalances the gain of microbial production. In the future, research using genetic technology may be able to produce a better bacteria-based plastic that has more desirable properties and is cheaper to produce.

5.5.5 Cellulose Graft-Polymers

The biopolymers cited above that are the product of ring-opening polymerisation reactions (PLA, PCL and PGA) can be involved in the synthesis of *graft*-polymers with polysaccharides (Teramoto and Nishio, 2004). As cellulose provides free hydroxyl groups, it can act as an initiator of the polymerisation (or copolymerisation) reaction. Depending on the operation conditions, oligomers of 3 to 12 units are formed as side chains of a main skeleton. The result is a biodegradable nanocomposite capable of forming alternated nanophases (Figure 5.26).

It is more convenient to substitute cellulose with partially substituted cellulose acetate (DS = 1). By this means, the main polymer is soluble in the lactone and the reaction is done in only one phase. The obtained products are more homogeneous and more resistant (Figure 5.27).

Cellulose can be replaced by starch or even proteins (Bhattacharya *et al.*, 1997). Of course, it is possible to carry out other *graft*-polymerisation reactions on cellulose or starch, for instance with acrylic monomers. The obtained material is useful, but it lacks biodegradability due to the synthetic side chains.

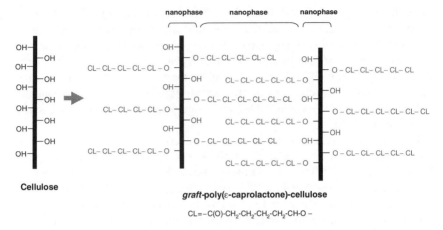

Figure 5.26   *Representation of the nanocomposite formed by graft-polymerisation of caprolactone on cellulose*

5.6   Conclusion

The world of biomaterials, as defined in this chapter, was introduced from a point of view considering essentially the technical feasibility. However, it can only be fully comprehended through a deep knowledge spread in numerous disciplines: agronomy, chemistry, biochemistry, economy, sociology, medicine, engineering and ecology.

Moreover, the development of these biomaterials will depend, not only on technical aspects, but also on political decisions and trends. Only the determination of industrialised societies to progress will be the motor for the promotion of ecologically friendly materials.

Figure 5.27   *Film obtained from graft-poly (ε-caprolactone)-cellulose acetate*

References

Anonymous. Wood-plastic composite growth taking off in Europe... while strong WPC growth continues in the USA.. *Additives for Polymers* (5):9–11 (2006).

Augier L., Sperone G., Vaca-Garcia C., Borredon E. Influence of the wood fibre filler on the internal recycling of PVC-based composites. Polym. Degrad. Stab. 92:1169–1176 (2007).

Banoub J.H., Benjelloun-Mlayah B., Ziarelli F., Joly N., Delmas M. Elucidation of the complex molecular structure of wheat straw lignin polymer by atmospheric pressure photoionization quadrupole time-of-flight tandem mass spectrometry. *Rapid Communications in Mass Spectrometry 21*:2867–2888 (2007).

Bhattacharya M., Jacob J., Vaidya U. Method of grafting functional groups to synthetic polymers for making biodegradable plastics. World patent (US) WO9747670 (1997).

Dong J., Sun Q., Wang J-Y. Basic study of corn protein, zein, as a biomaterial in tissue engineering, surface morphology and biocompatibility. *Biomaterials 25*:4691–4697 (2004).

Dulbecco P., Luro D. L'essentiel sur le bois. Centre Technique du Bois et de l'Ameublement. Paris, pp.184 (2001).

Dumonceaud O., Thomas R. Process and apparatus for wood impregnation. French patent FR 2870773 (2004) and www.oleobois.com

Ekman K., Eklund V., Fors J., Huttunen J., Mandell L., Selin J.F. Turunen O.T. Regenerated cellulose fibers from cellulose carbamate solutions. *Lezinger Berichte 57*:38–40 (1984).

El Seoud O., Heinze T. Organic esters of cellulose: New perspectives for old polymers. *Advances in Polymer Science 186*:103–149

Forss K., Fremer K-E. The nature of lignin: a different view. In *Lignin: Historical, Biological, and Materials Perspectives*. Glasser W.G., Northey R.A., and Schultz T.P. (eds). ACS Symposium Series 742, p.100–116. (2000)

Funaoka M. Novel lignin derivatives, molded products using the same and processes for making the same. American patent. US2005154194 (2005).

Gaspar M., Benko Zs. Dogossy G. Reczey K., Czigany T. Reducing water absorption in compostable starch-based plastics. *Polymer Degradation and Stability 90*:563–569

Glasser W., McCartney B., Samaranayake G. Cellulose derivatives with low degree of substitution. 3. The biodegradability of cellulose esters using a simple enzyme assay. *Biotechnology Progress 10*:214–219 (1994).

Hamalainen C., Wade R., Buras E.M. Jr. Fibrous cellulose esters by trifluoroacetic anhydride method. *Text. Res. J. 27*:168–168 (1957).

Hill C.A.S., Jones D. The dimensional stabilization of Corsican pine sapwood due to chemical modifications with linear chain anhydrides. *Holzforschung 50*:457–462 (1996).

Kokta, B.V., Maldas D., Daneault C., and Beland P. Composites of poly(vinyl chloride)-wood fibers. I. Effect of isocyanate as a bonding agent. *Polymer-Plastics Technology and Engineering* 29(1 – 2):87–118 (1990).

Kokta, B.V., Maldas D., Daneault C., Beland P. Composites of poly(vinyl chloride) - wood fibers. III: Effect of silane as coupling agent. *Journal of Vinyl Technology 12(3)*:146–53 (1990).

Kwatra, H., Caruthers, J., Tao, B. Synthesis of long chain fatty acids esterified onto cellulose via the vacuum-acid chloride process. *Ind. Eng. Chem. Res. 31*:2647–2651 (1992).

Lin L., Nakagame S., Yao Y., Yoshioka M., Shiraishi N. Liquefaction mechanism of β-O-4 lignin model compound in the presence of phenol under acid catalysis. Part 2. Reaction behavious and pathways. *Holzforschung 55*:625–630 (2001).

Magne M., El Kasmi S., Dupire M., Morard M., Vaca-Garcia C., Thiebaud S., Peydecastaing J., Borredon M.E., Gaset A. Procédé de traitement de matières ligno-cellulosiques, notamment du bois ainsi qu'un matériau obtenu par ce procédé. French patent FR 2838369 (2003) and www.woodprotect.fr

Morard M., Vaca-Garcia C., Stevens M., Van Acker J., Pignolet O., Borredon E. Durability improvement of wood by treatment with Methyl Alkenoate Succinic Anhydrides (M-ASA) of vegetable origin. *International Biodeterioration & Biodegradation 59*:103–110 (2007) and www.surfasam.com

Park I.M., Ibáñez A.M., Shoemaker C.F. Rice Starch Molecular Size and its Relationship with Amylose Content. *Starch/Stärke 59*:69–77 (2007).

Rouilly A., Jorda J., Rigal L. Thermo-mechanical processing of sugar beet pulp. I. Twin-screw extrusion process. *Carbohydrate Polymers 66*:81–87 (2006).

Rouilly A., Orliac O., Silvestre F., Rigal L. New natural injection-moldable composite material from sunflower oilcake. *Bioresource Technology 97*:553–561 (2005).

Rowell R.M. Chemical modification of agro-resources for property enhancement. In *Paper and Composites from Agro-Based Ressources*. R.M. Rowell, R.A. Young and J.K. Rowell (eds.). Lewis Publishers CRC., pp. 351–375 (1997).

Rowell R.M., Keany F.M. Fiberboards made from acetylated bagasse fiber. *Wood Fiber Sci. 23*:15–22 (1991).

Samaranayake, G. Glasser, W. Cellulose derivatives with low DS. I A novel acylation system. *Carbohydrate Polymers 22*:1–7 (1993).

Shiraishi N., Aoki T., Norimoto M., Okumara M. Make cellulosics thermoplastic. *Chemtech (June)*:366–373 (1983).

Silvestre F., Rigal L., Leyris J., Gaset A. Aqueous adhesive based on a vegetable protein extract and its preparation. European patent EP997513 (2000).

SK Corp. Biodegradable linear low-density polyethylene composition and film thereof. Korean patent KR20010084444 (2001).

Teramoto Y., Nishio Y. Structural designing and functionalization of biodegradable cellulosic graft copolymers. *Cellulose Communications 11*:115–120 (2004).

Thiebaud, S., Borredon, M.E. Solvent-free wood esterification with fatty acid chlorides. *Bioresource Technology 52*:169–173 (1995).

Tokiwa Y., Takagi S., Koyama M. Starch-containing biodegradable plastic and method of producing same. American patent US5256711 (1993).

Vaca-Garcia C., Gozzelino G., Glasser W.G., Borredon M.E. DMTA transitions of partially and fully substituted cellulose fatty esters. *Journal of Polymer Science. Part B. 41*:281–288 (2003).

www.accoya.info
www.retiwood.com
www.solstitch.net
www.wtt.dk

Yang I., Kuo M., Myers D., Pu A. Comparison of protein-based adhesive resins for wood composites. *Journal of Wood Science 52*:503–508 (2006).

6

Production of Energy from Biomass

Mehrdad Arshadi[a] and Anita Sellstedt[b]

[a] Swedish University of Agricultural Sciences (SLU), Sweden [b] Umeå University, Sweden

6.1 Introduction

The transport sector generates a lot of emissions of different gases and particles in the air. The gases are released by combustion of different fossil fuels in car engines. These emissions have a negative effect on human health and are also now known to cause vast environmental changes. Several of these gases are greenhouse gases such as CO_2, CH_4, N_2O, etc. There is very strong evidence of a correlation between greenhouse gases and worldwide weather changes resulting in global warming. These negative environmental effects together with increasing energy demands in the world and depletion of fossil fuels in the near future, in combination with long-distance transport have had a positive impact on developing new energy sources.

Bioenergy (energy derived from biological sources) production from renewable resources has increased during recent decades. World energy demand has dramatically increased and new sources of energy, beside fossil resources, are needed. The European Union is prompting its member states to increase their use of biofuels to meet a target share of 5.75% of the total fuel market by 2010. However, they had only reached an average share of 1.4% of the biofuel usage for all member states in 2005. To meet the EU target it is necessary to act in several different areas and develop several different bioenergy products, depending on regional possibilities

Introduction to Chemicals from Biomass Edited by James Clark and Fabien Deswarte
© 2008 John Wiley & Sons, Ltd

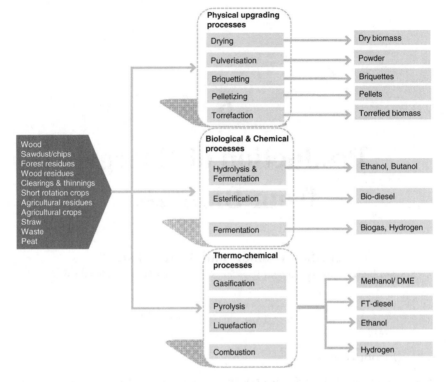

Figure 6.1   Energy products and classification

and the availability of natural resources. In addition, it is important to see if there is any pre-existing infrastructure that can be used. Furthermore, it is important to consider economic, environmental, national security and technological factors in all of the different countries.

In general, energy sources are categorised as solid (e.g. wood, pellets), liquid (e.g. ethanol, biodiesel) and gas (e.g. biogas, hydrogen) fuels. But also, depending on the kind of process used in the production of the energy products, they can be divided in four groups, i.e. physical upgrading, microbiological, thermochemical and chemical processes. Energy products and classification are shown in Figure 6.1.

6.2   Physical Upgrading Processes

6.2.1   Refinement of Solids to Biofuel

The production of solid fuels from renewable resources (e.g. biomass) has become more important due to increasing energy demands and also because of environmental pressure. Upgraded biofuels (refined biofuel) include powdered fuels and densified fuel such as briquettes and pellets. The raw materials are often shavings and sawdust from sawmills. The interest in new biological raw materials, such as grasses, and agricultural residues, such as corn stover, olive seeds, wheat straw, peat, etc. has increased recently (Olsson, 2006).

6.2.2 Wood Powder

Wood powder is a kind of upgraded fuel that is burned in large-scale combustion plants for heat production. However, it is possible to use wood powder for power generation as well. It is a biofuel made of sawdust, shavings and bark. The raw material is crushed, dried and milled to fine particles in order to obtain the best fuel properties. There are many different wood powder qualities dependant on different physical properties such as particle size distribution, particle shape and also moisture content. The powder is usually handled in a closed system from milling to storing in silos to avoid the risk of dust explosions. The raw materials and type of mill used determine the properties of the wood powder (Paulrud *et al.*, 2002).

6.2.3 Briquette Production

Briquettes are another form of upgraded biofuel. Agricultural and forestry residues and other biological materials are often difficult to use as biofuel because of their inhomogeneous, uneven, bulky and troublesome characteristics. This problem can be solved by densification of the residues into compact, regular shapes. The technology is based on forcing the biomass under high pressure, by means of a screw or piston-press, through a die. The friction between the raw material and equipment increases the temperature inside the press, with lignin usually acting as the binder. The diameter of the briquettes can be between 35 and 90 mm. There are many advantages to this process, such as:

- Increasing the net calorific content of the material per unit volume; briquettes with a density of about 800–1300 kg m^{-3} are usual, compared to loose biomass with a bulk density of 10–20 kg m^{-3} or to the bulk density of wood chips at about 300 kg m^{-3}.
- A resultant product of uniform size and well-defined quality and moisture content.
- A product that is easy and cheap to handle, transport and store.
- Easier optimisation of combustion, resulting in higher efficiency, lower emissions and a smaller amount of ash.
- Considerably lower investment in furnaces and purification equipment.

Briquetting of biomass can be achieved by different techniques, and by either using some kind of binder or by direct compacting without any binder. One example from a pilot plant is presented below. Bales of raw material are put on a conveyor, transported to a shredder and stored in a silo. The shredder cuts the material into particles ranging from microscopic to 15 mm. When the production of briquettes starts, material from the silos is transported through a mixer and a separator to a buffer silo above the briquette press. At the bottom of the buffer silo there is an auger, which transports the material to the press. The briquette press has a high production capacity, e.g. 450–500 kg h^{-1}. A haymatic H_2O tester can be used to get an estimation of the humidity of the raw materials. This instrument measures the electrical conductivity of the raw materials and gives a rough estimation of the moisture content at the surface of a sample. It is also possible to measure the moisture content of the raw material continually by on-line near-infrared spectroscopy (NIR spectroscopy) in combination with multivariate data analyses. Several parameters affect the mechanical strength of the briquettes, such as the moisture content of the raw material, pressure and temperature. Several other raw materials have been used in the briquette process, such as the dry fraction of household waste as an admixture to other cellulose material, like straw and reed canary grass. Briquettes show good fuel qualities with regard to combustion properties and the addition of waste to the biofuel does not cause any increase in emissions of organic compounds during combustion and heat production (Hedman *et al.*, 2005).

6.2.4   Pellet Production

A wood pellet is a small, hard piece of bioenergy. Normally pellets have a cylindrical form, 6–8 mm in diameter and of varying length. In the last decade, softwood pellets have emerged as a renewable energy resource that can be considered as a potential future substitute, in many aspects, for fossil fuels such as oil and natural gas. The energy value of 1 ton of pellets is about 5.0 MWh, which is equal to 0.5 m^3 oil.

Since pellets have a closed CO_2 emission circle, their usage is one step forward to a sustainable energy system. Wood pellets have been used in the household sector as well as the industrial sector as an energy resource for heat and power generation. Today, there are many countries which produce wood pellets and also some which export the fuel overseas by ship, e.g. from Canada to Europe. The raw material is a forest by-product, such as sawdust, mostly from pine and spruce, and also cutter shavings and bark of those species. However, there is more interest in new raw materials for pellet production, since the lignocellulose material from softwoods is also used in the pulp and paper industry and recently, cellulose in ethanol production. Therefore, other biomass such as straw, different grasses, corn stover and other agricultural products are valuable candidates for pellet production (Mani, 2006). It is possible to use a mixture of different lignocellulose materials in pellet production and there is also ongoing research in Sweden and China on pelletising corn stover, rape seed cakes, eucalyptus leaves, bark, peat, hemp and cotton.

In ethanol production from cellulose material, there is a huge amount of lignin left as a by-product. The lignin is a valuable raw material for chemical production in a biorefinery. So in order to reduce the transport cost of the lignin residue, it may be possible to mix it with other raw materials in pellet production in the near future.

The first generation pellet presses were developed for feed production, but over several years of experiments for wood pellet production, several modifications have occurred and the new wood pellet equipment is now well optimised. In a traditional wood pellet production process the raw material is dried to about 5–15% moisture content, which requires a lot of energy. In the next step, the dried material is ground into a fine powder (less than 3 mm in diameter). The powder is then pressed through cylindrical holes in a pellet matrix (die) to make short sticks, which are called pellets. The pressure inside the pellet press is very high (about 70–100 MPa). In order to improve the production capacity, steam can be added before pressing to adjust the raw material moisture content and some softening of lignin/hemicelluloses occurs, which improves particle binding. During pelletising, the temperature increases to more than 120 °C. The high temperature causes the wood components, such as lignin, to soften and thus helps to keep the particles together as wood pellets. Therefore, usually no binder is needed. Generally, binders increase production costs and also have a negative impact on combustion properties. The produced pellets must be cooled to ambient temperature relatively rapidly to prevent water absorption from moisture in the surrounding air. One drawback of the pelletising technology is its high consumption of electricity during the production process. Another disadvantage of pellets is the relatively high amount of fines and dust produced during the pelletising process and during storage. This presence of fines and dust increases the production cost of pellets, since both must be removed (e.g. to avoid dust explosions) before packaging and delivery of the pellets. It is also important to remove any possible metals, stones or sand that may contaminate the sawdust before the pelletising process begins. Figure 6.2 shows schematically the different steps in a pilot plant for pellet production.

In pellet production there is a complex interaction of forces between the particles (Mani, 2006). Once manufactured, the pellets are in an optimal form for extended storage and transport without any loss of quality compared to sawdust, which has a moisture content of 45–55% and needs a lot of space to be stored. In addition, sawdust is changed by microbial and chemical activities during storage.

Pellets are a cylindrical form of densified wood particles that have several advantages compared to the raw material (mostly sawdust):

- Lower moisture content (about 40–45% less water content)
- Better homogeneity which results in less variation in moisture content and particle size and better combustion properties
- Less uneven combustion with undesirable emissions
- Higher density and lower transport costs
- Higher energy density and much easier feeding of burners
- Lower storage costs
- The possibility of longer storage times without any risk of mould formation and other microbiological and chemical activities.

Process control in pelletising production can be aided by on-line near-infrared spectroscopy (NIR). The NIR technique has a high potential for process control

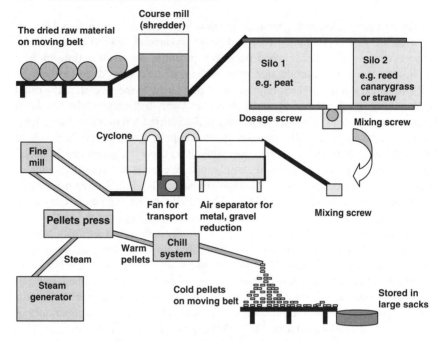

Figure 6.2   Scheme of a pilot plant for pellet production

in many steps of pellet production, e.g. it is possible to determine the moisture content of raw material and pellets at any time in the process very quickly and with high accuracy.

The produced pellets are usually stored in closed warehouses or silos for protection against water absorption. During long-term storage or transportation of wood pellets, some emissions of volatile organic compounds, such as aldehydes and low molecular carboxylic acids, may be released in the storage room or in the containers, causing unpleasant smells and possibly causing health problems. These organic compounds are produced by oxidation of fatty and resin acids in the wood pellets (Arshadi and Gref, 2005).

The pellets are then delivered in bulk by lorry, train or boat and also in large or small sacks to consumers. Pellets are burned in pellet furnaces to produce heat in small-, medium- and large-scale facilities. Since the quality of pellets may vary during production there are some parameters that can be used to evaluate this. The most important physical parameters are bulk density, pellet density, durability, which can be measured by determining the amount of fines, and particle size distribution of the raw material. Chemical parameters include pellet moisture content, calorific heat value as a determinant of energy content, element content (such as K, Mg, Ca, N, S, Cl), ash content and ash melting behaviour in order to prevent the agglomeration of the ash in the furnace. The net calorific heat value of pellets is about 19 MJ kg^{-1}, which is less than 50% of the oil equivalent. Pellets consist of water, a flammable part and also ash. The ash content from pellet burning is usually less than 0.5% of the gross weight. Ash contains several elements, such as phosphorus, potassium, calcium, magnesium and silicon. A high amount of fines (sawdust released from pellets, for example during transportation) in pellets causes more ash and other problems associated with their combustion. In the worst case, ash can melt in the burner and destroy the furnace. When pellet combustion is not optimised, the emissions are CO, caused by insufficient oxygen or low combustion temperature, NO_x, often caused by too high a combustion temperature (the origin of N is from pellets and/or air), CO_2, total HC and particles. The amount of chlorine and sulfur in pellets is also another important factor in the reduction of HCl, dioxin and SO_x emissions during combustion.

6.2.5   Torrefaction

Torrefaction of woody biomass is an upgrading process for drying of biomass, removing some volatile organic compounds and decomposing the reactive hemicellulose fraction of the wood. Torrefaction is a mild pyrolysis process, which improves some properties of wood. The process increases the carbon content and net calorific value compared to untreated wood. In the process, the woody material is heated in the absence of oxygen to about 225–300 °C for several minutes. In this temperature range the decomposition of a part of the hemicellulose occurs. This process will become important in the future as a pre-treatment for increasing energy density before gasification of biomass (Prins, 2006).

6.3   Microbiological Processes

6.3.1   Organisms and Processes

Bioenergy is defined as a source of energy that is renewable, as opposed to a fossil fuel such as oil, which is a non-renewable source. The material used for bioenergy has a biological origin, i.e. it is a biological organism that has produced it. This biological organism or agent can be bacteria, algae or plants. Algae, plants and some bacteria have one thing in common, they are autotrophic, that is, they can use the energy of the sunlight directly in a process called photosynthesis. There are also bacteria that do not use this process, that is, they are heterotrophic. In photosynthesis the light energy from the sun is captured in the biological machinery and the carbon dioxide from the air is built into carbohydrates that can be used by the organism to grow and/or to make cellulose, i.e. biomass production. A very large part of the world's biomass consists of plant cell walls. On burning the material for bioenergy, carbon dioxide, oxygen, water and ash are the by-products. The photosynthetic process is quite similar in plants from different ecosystems, such as forests, agricultural lands and mires. There is, however, a difference in photosynthesis between temperate species, e.g. pine, and species adapted to sub-tropical and tropical climates, e.g. maize, with the former having C-3 photosynthesis and the latter having C-4 photosynthesis. Biomass is considered to be a renewable source if it is handled in the right way. However, if forests and agricultural lands are harvested without thinking of renewal, then that harvesting can cause erosion, and the product cannot be considered renewable. World energy use is dominated by fossil fuels, that is, coal, oil and gas. Fossil fuels have taken millions of years to be produced, while renewable biomass is produced within the time frame of 100 years. That it is why it is debated whether peat is a biofuel or a fossil fuel, because it only takes around 10000 years to be formed. In order to utilise the energy that has been trapped by photosynthesis there are a variety of microbial activities that can be used, and some of them are described below.

6.3.2   Microbiological Ethanol Production

The production of ethanol is a technique with a history of thousands of years. The ancient Greek society already produced ethanol and used agricultural products such as grains as substrate. Recently biomass in the form of woody material has been discussed as having a great future potential.

Feedstock for Microbiological Ethanol Production

Lignocellulosic or woody biomass is composed of carbohydrate polymers such as cellulose, hemicellulose, lignin and to a much smaller degree, extractives, acids, salts and minerals. Cellulose and hemicellulose can be hydrolysed to sugars and fermented to ethanol. Cellulose is a polymer of glucose – glucose dimers, and it is the linkages and presence of hydrogen bonds that make the cellulose polymer difficult to break. The process of saccharification involves the addition of water to break the linkages and getting the product glucose-free for ethanol production.

Hemicellulose consists of short, highly branched chains of different sugars, such as the five-carbon sugars xylose and arabinose, but also the six-carbon sugars mannose, glucose and galactose. Because of the branched structure of hemicellulose it is relatively easy to break for fermentation processes. Lignin is one of the most resistant materials on Earth and cannot be used in the fermentation processes, but lignin is present in all biomass to a variable extent, depending on the origin of the biomass. Both grain and woody material are examples of biomass, although they can vary somewhat in composition. As an example, switch grass contains 31.98% cellulose, 25.19% hemicellulose and 18.13% lignin (Hammelinck *et al.*, 2005) while pine contains 44.55% cellulose, 25.19% hemicellulose and 27.67% lignin (Hammelinck *et al.*, 2005) as is shown in Table 6.1. Pine biomass has a slightly higher content of sugars in the biomass, thus making it valuable for ethanol production. However, irrespective of whether pine or spruce is the source in the ethanol fermentation processes, inhibitors are produced. In addition, pine has a large content of lignin, which makes it suitable to be used for pellet production, as opposed to grasses, which have a lower content of lignin (Hammelinck *et al.*, 2005). Thus, glucose and xylose are the main fermentable sugars in

Table 6.1 *Typical lignocellulosic biomass compositions[a,b] (% dry basis). (Reprinted from Hamelinck et al. (2005), with permission from Elsevier.)*

Feedstock	Hardwood			Softwood	Grass
	Black locust	Hybrid Poplar	Eucalyptus	Pine	Switch grass
Cellulose	41.61	44.70	49.50	44.55	31.98
Glucan 6C	41.61	44.70	49.50	44.55	31.98
Hemicellulose	17.66	18.55	13.07	21.90	25.19
Xylan 5C	13.86	14.56	10.73	6.30	21.09
Arabinan 5C	0.94	0.82	0.31	1.60	2.84
Galactan 6C	0.93	0.97	0.76	2.56	0.95
Mannan 6C	1.92	2.20	1.27	11.43	0.30
Lignin[c]	26.70	26.44	27.71	27.67	18.13
Ash	2.15	1.71	1.26	0.32	5.95
Acids	4.57	1.48	4.19	2.67	1.21
Extractives[d]	7.31	7.12	4.27	2.88	17.54
Heating value[e] (GJ_{HHV} $tonne_{dry}^{-1}$)	19.5	19.6	19.5	19.6	18.6

[a] The exact biochemical composition of biomass depends on many different factors, such as growth area, fertilizers used, time of harvesting and storage conditions. Softwood hemicellulose yields more 6C sugars, whereas hardwood yields more 5C sugars.
[b] From database at US DOE Biofuels website. The fractions from the source data have been corrected to yield 100% mass closure.
[c] Bark and bark residues have a relatively higher lignin content. The empirical formula for lignin is $C_9H_{10}O_2(OCH_3)_n$, with n the ratio of MeO to C9 groups: $n = 1.4$, 0.94 and 1.18 for hardwood, softwood and grasses, respectively.
[d] Low molecular weight organic materials (aromatics, terpenes, alcohols), some of which may be toxic to ethanol fermenting organisms, and cause deposits in some pretreatments. Some compounds could be sold as chemicals (e.g. antioxidants) having a higher value than ethanol, but costs for purification are unknown .
[e] However, a relation between biomass composition and heating values has been suggested by various authors, the higher heating value of lignin is 24.4 ± 1.2 GJ $tonne_{dry}^{-1}$, whereas the holocellulose plus the rest have a heating value of about 17 GJ $tonne_{dry}^{-1}$.

all lignocellulosic material, ranging from rice to hardwood, with the exception of softwoods, which also contain a considerable amount of the six-carbon sugar mannose. This difference in composition puts different requirements on the production process.

Microbiological Ethanol Process

The production of ethanol requires pretreatment of the biomass before fermentation. The pretreatment can be performed physically, chemically or biologically. Physical pretreatments include communition (dry, wet and vibratory ball milling), irradiation (electron beam irradiation, or microwave heating), steaming or hydrothermolysis (Hsu, 1996). Chemical pretreatments include treatments with acids, alkaline substances, solvents, ammonia, sulfur dioxide or other chemicals. Biological pretreatments comprise techniques where decomposing microorganisms or enzymes are added. In this way, biological pretreatments are more environmentally friendly, but, to date, have been found to be slower and therefore less economically friendly. Interestingly, the use of enzymes to pretreat cellulose is believed to render a higher ethanol yield.

The production of ethanol uses the glycolysis pathway, a common biochemical pathway that is present in animals as well as in yeast and bacteria. In this pathway glucose is commonly converted to pyruvate when oxygen is present (aerobiosis), but if oxygen is lacking (anaerobiosis), ethanol is produced.

The overall reaction when six-carbons are converted to ethanol is

$$C_6H_{12}O_6 \longrightarrow 2\,C_2H_5OH + 2\,CO_2$$

while the reaction with five-carbon sugar is

$$3\,C_5H_{10}O_5 \longrightarrow 5\,C_2H_5OH + 5\,CO_2$$

Ethanol production is a technique where microorganisms such as *Saccharomyces cerevisae, Zymomonas* sp. and *Candida* sp. are the fermentation agents. It is, however, the well-known fungi, *Saccharomyces cerevisae*, baker's yeast, that is most commonly used. *S. cerevisae* can be used as a natural organism, but also as a genetically modified one, a so-called recombinant. Interestingly, several authors have reported on the gene modification of *S. cerevisae* using different techniques (reviewed by Sedlak and Ho, 2004). It was recently acknowledged that cloning three genes into *S. cerevisae* is needed to make it really efficient for fermentation (Sedlak and Ho, 2004).

Also of great interest is a recently discovered fungal mix that has been patented (Patent WO-2005054487). The patent describes a novel mix of fungi that can ferment both six-carbon and five-carbon sugars to ethanol. In addition, the patent describes the fact that the fungi increase ethanol production when used together with *S. cerevisae* and, thus, can be used in already existing industrial set-ups.

Enzymatic Hydrolysis

Irrespective of the type of biomass used for ethanol production, the biomass needs to be pretreated to make the carbohydrates available for fermentation. However, which enzymes can be used depends on the source of the biomass. In addition, the biomass needs pretreatment before the enzymes are used. The first step of the pretreatment can be of a physical nature. Once the biomass is physically pretreated, the cellulose structures are open for enzyme action. In biomass from forests, the substance is mainly in the form of cellulose. Targeted enzymes are selective for the reaction of cellulose to glucose, and therefore there are no degradation by-products, as occurs in acid conversion technology. There are at least three ways this can be performed. Firstly, in separate hydrolysis and fermentation, the pretreated biomass is treated with cellulase, which hydrolyzes the cellulose to glucose at 50 °C and pH 4.8. Secondly, in simultaneous fermentation and saccharification (SSF) the hydrolysis and fermentation take place in the same bioreactor. Thirdly, in direct microbial conversion (DMC) the microorganisms produce the cellulase as well as producing ethanol. As an example, the Canadian company Iogen Ltd uses especially efficient celllulases that are produced at Iogen Ltd, after a modified steam explosion as the pretreatment. For other sources of biomass, however, other enzymes are needed. For starch-rich biomass there is a need for α-amylase to convert the starch to glucose monomers. Functionally, cellulases can be produced by a variety of microorganisms, but it is a species of *Trichoderma* that is considered to be the most efficient. This fungi produces a complex mixture of cellulases consisting of a variety of different enzymes. The actions of the cellulases are synergistic, i.e. the sum of their actions is greater than the individual action of each of them.

6.3.3 Production of Butanol From Bacteria

Several bacteria, such as *Clostridium beijerinckii*, are capable of converting carbohydrates to butanol. The bacteria performing this conversion are all heterotrophic, i.e. they need already made carbohydrates as a substrate for the production of butanol.

6.3.4 Production of Biodiesel From Plants and Algae

Biodiesel is a fuel derived from plant oils or algae. Biodiesel is a biodegradable and non-toxic diesel equivalent. A variety of feedstocks can be used for biodiesel production e.g. rapeseed and soybean oils, where soybean oils are most commonly used. In addition, animal fats including yellow grease and chicken fat can be used and also sewage waste as a substrate for algae, which themselves produce biodiesel. The feedstock that produces the highest yield is algae (Chisti, 2007), being several times more efficient than soybean.

The components of biodiesel are vegetable oils composed of glycerol esters of fatty acids. In the process of transesterification, the glycerol components of the triglyceride molecules are exchanged for methanol. The products are fatty-acid methyl esters consisting of straight saturated and unsaturated hydrocarbon chains, as described under chemical processes.

The biological production is described in the reaction formula

Sunlight + carbon dioxide $\rightarrow$ algal biomass $\rightarrow$ biodiesel + spent biomass

Irrespective of biodiesel feedstock, the photosynthetic process is used to convert solar energy into chemical energy. The chemical energy is stored mainly as carbohydrates, proteins and fatty acids. Most of the carbon dioxide emitted when burning biodiesel is simply recycled carbon dioxide that was absorbed during plant growth, so the net production of greenhouse gases is small. The production of biodiesel from algae widely exceeds the amount that is produced from soybean, for example, as is described in Table 6.2 above.

Table 6.2   *Comparison of some sources of biodiesel. (Rewritten after Cristi, 2007)*

Crop	Oil yield (L/ha)	Land area needed (Mha)[a]	Percentage of existing US cropping area[a]
Corn	172	1540	846
Soybean	446	594	326
Microalgae[b]	136900	2	1.1
Microalgae[c]	58700	4.5	2.5

[a] For meeting 50% of all transport fuel needs of USA
[b] 70% oil (by wt) in biomass
[c] 30% oil (by wt) in biomass

6.3.5 Biogas Production

Naturally occurring ecosystems such as waterlogged swamps, bogs and marshes are inhabited by microorganisms that have evolved over millions of years under oxygen-limiting to oxygen-free conditions. The microorganisms are capable of utilising organic and inorganic substrates for their own metabolic activities. Biogas production, i.e. production of methane and carbon dioxide with trace amounts of hydrogen gas, nitrogen gas and hydrogen sulfide, is thus a naturally occurring process in which humankind has lately developed an increasing interest.

The reaction formula can be written as follows:

Complex organic materials $\rightarrow$ Monomeric organic compounds $\rightarrow$

Intermediate products $\rightarrow$ H_2 + CO_2 + acetate $\rightarrow$ CH_4 + CO_2

The process is described in Figure 6.3.

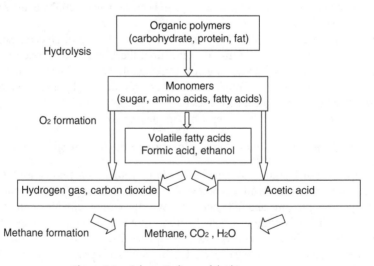

Figure 6.3   *Schematic figure of the biogas process*

The first part of the process is the degradation of complex organic compounds (lipids, proteins, carbohydrates) to monomers and oligomers (fatty acids, amino acids, monomers of carbohydrates), a process called hydrolysis. This part of the process is the most rate-limiting step of the anaerobic production of methane and after this, fermentation can take place, producing intermediary compounds such as alcohols, long fatty acids and acetate. The anaerobic oxidation yields hydrogen gas, carbon dioxide and acetate. The final part of the process, methanogenesis, involves conversion of the hydrogen, acetate and carbon dioxide to mainly methane and carbon dioxide. There are, however, by-products, such as hydrogen gas and nitrogen gas.

Microorganisms Involved in Biogas Production

A large variety of morphologically different methanogenic bacteria have been isolated ranging from rod-shaped, cocci and plate-shaped to filamentous types. However, molecular techniques, such as the analysis of 16S rRNA sequences, have revealed the real identity of the methanogenes. To date, seven different groups of methanogenes have been described, including species of *Methanobacterium, Methanococcus, Methanosarcina* and *Methanopyrus*. The most common substrates for methane production for these bacteria are hydrogen gas, carbon dioxide and formate.

The reactions leading to the formation of methane involve several microorganisms in the anaerobic chain that depend on cooperation to get to the right end product. The microorganisms at the end of the reaction chain are very dependent on those at the beginning. In some reactions the bacteria involved have adapted totally to a life in collaboration. In the Figure 6.3 there are the hydrogenotrophic methanogenes e.g. *Methanobacteriaceae* sp. Interestingly, another group of bacteria, the acetotrophic methanogens (e.g. *Methanosarcina* sp.) are involved, as shown in Figure 6.3. Interestingly, hydrogen (see next section) is produced in several steps of the methane pathway, indicating that these two reactions naturally interact in nature (Robson, 2001).

How to Enhance Biogas Production

Scientists have proposed a variety of ways to enhance biogas production as reviewed by Yadvika *et al.* (2004). The different ways of enhancing biogas production involve the use of additives, gas enhancement by recycling of digested slurry, variation of operational parameters and the use of fixed films. We will only briefly consider additives and temperature and C/N ratio as operational parameters. Additives are mainly used in order to increase microbial activity and to help maintain favourable conditions with regard to pH, lowered inhibition of acetogenesis, and methanogenesis. The parameter that has the greatest impact on biogas production is temperature. Biogas can be produced in the temperature range <30 °C (psychrophilic), 30–40 °C (mesophilic) and 40–50 °C (thermophilic). Anaerobic bacteria are generally active in mesophilic and thermophilic temperature ranges, so these temperatures yield more biogas. Production can take place in the psychrophilic range as well, but a lot more time is required. It is also important to maintain the right C/N ratio in the substrate. The anaerobic bacteria active in biogas production have been shown to use carbon 25–30 times faster than nitrogen (Yadvika *et al.*, 2004), and therefore the C/N ration needs to be 20–30:1 to get optimal conditions for the bacteria.

6.3.6 Hydrogen Production

Hydrogen is the most abundant naturally occurring element in the universe and, in addition, a very efficient energy carrier. This extremely small molecule is metabolised by a variety of organisms ranging from bacteria, archaebacteria and cyanobacteria to green algae. The simple conversion of protons to hydrogen gas, which in most organisms is a reversible reaction, has led to the evolution of a large number of enzymes (so called hydrogenases) with a variety of subunits to achieve the task. Thirteen different hydrogenases have been identified (Robson, 2001), most of them dealing with energy reactions. There are excellent reviews on the biodiversity and functions of hydrogen-producing organisms (e.g. Adams, 1990; Robson, 2001), so only some examples will be mentioned in this chapter.

The Process of Hydrogen Production

$$2\,H^+ \rightleftharpoons H_2 + 2\,e^-$$

As is indicated in the reaction formula, the process is reversible, so hydrogen is either produced or consumed. The microorganisms involved in hydrogen production are either those capable of photosynthesis (phototrophic) or those that do not have photosynthesis (heterotrophic organisms). The phototrophic organisms can be divided into photoautotrophic and photoheterotrophic groups. The phototropic organisms are represented mainly by microalgae and cyanobacteria and are able to use light as an energy source and carbon dioxide as carbon source. The photoheterotrophs are cyanobacteria that in spite of the fact that they actually can use light energy and carbon dioxide also need organic carbon for their nitrogen fixation. Taking into account the way of getting carbon, the reaction formulas will of course be more complicated.

Photoautotrophic Hydrogen Production

Examples of photoautotrophic organisms are algae such as *Scenedesmus* sp. and *Chlamydomonas reinhardtii* (Melis *et al.*, 2000). Interestingly, Professor Melis *et al.* (2000) have developed a two-stage photosynthesis and hydrogen production system with the green algae *C. reinhardtii*. It has been shown that a lack of sulfur causes a specific, but reversible, decline in the rate of the oxygen-producing step of photosynthesis. In enclosed cultures, this imbalance in the photosynthesis–respiration relationship caused by S-deficiency results in a net consumption of oxygen, in turn, resulting in anaerobiosis in the culture and an increased hydrogen production. The two-step system allows the culture to grow and have a normal photosynthesis apparatus in a complete growth medium, with the medium then being switched to S-deficiency and hydrogen evolution beginning.

Photoheterotrophic Production of Hydrogen

The photoheterotrophic microorganisms are represented by cyanobacteria such as *Nostoc* sp. and *Anabaena variabilis*. They are capable of photosynthesis using sunlight and carbon dioxide in a similar manner as the photoautotrophs. They are also capable of nitrogen fixation mediated by a nitrogenase enzyme, and this is where the carbon is needed, according to the formula

$$N_2 + 8\,H^+ + 8\,e^- + 16\,ATP \longrightarrow 2\,NH_3 + H_2 + 16\,ADP + 16\,Pi$$

Hydrogen production from acetate demands energy that is gained from light according to the formula

$$C_2H_4O_2 + 2\,H_2O + \text{light energy} \longrightarrow 2\,CO_2\,\text{gas} + 4\,H_2\,\text{gas}$$

Thus, these organisms have two enzymes involved in hydrogen production, nitrogenase and hydrogenase. There is no evidence that cyanobacterial strains have several hydrogen-producing enzymes, but there is a variety of strains yet to be evaluated for hydrogen production and the possibility exists to genetically modify them (Tamagnini *et al.*, 2002).

Heterotrophic Production of Hydrogen

$$C_6H_{12}O_6 + 4\,H_2O \longrightarrow 2\,CH_3COO^- + 2\,HCO_3^- + 4\,H_2$$

The equation above is based on the use of a six-carbon compound, while for natural conditions a larger variety of compounds is used. Heterotrophic bacteria, e.g. *Clostridium* sp., *Enterobacter* sp. *Ralstonia* sp., *Rhodobacter* sp. and *Frankia* sp., cannot use sunlight and therefore they need this supply of carbohydrates for their production of hydrogen. These organisms are naturally occurring in ecosystems all over the world. Processes with industrial applications have adopted natural systems and tried to mimic these for hydrogen production. One strain of *Frankia*, designated R43, was shown to produce a large quantity of hydrogen in anaerobic conditions using the carbon source propionate (Mohapatra *et al.* 2004). Figure 6.4 presents the different cells of *Frankia R43*, both in the hyphae and vesicles. It was earlier shown that the hydrogen-producing enzyme is present both in hyphae and vesicles (Mohapatra *et al.*, 2004).

With regard to the need for substrates for industrial hydrogen production, there is a wide range of substrates produced in different industries that could be used.

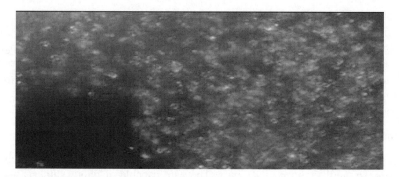

Figure 6.4   *Light micrograph of Frankia R43, showing spherical structures representing vesicles. (Light micrograph taken by Kjell Olofsson and A. Sellstedt, UPSC, SLU and Umeå University, Umeå, Sweden)*

Moreover, the conversion of carbohydrates to hydrogen is to be preferred, because it is thermodynamically superior. Glucose is a carbohydrate that is present in most industrial flows. Bioconversion of 1 mol of glucose theoretically yields 12 mol of hydrogen gas, if water is taken into account. Taking into account the reaction stoichiometry, bioconversion of 1 mol of glucose to acetate results in 4 moles of hydrogen gas per mol of glucose.

Heterotrophic Production of Hydrogen Coupled to Photoproduction

Bioconversion of biomass by use of heterotrophic bacteria yields hydrogen gas and most often acetate, because the oxidisation is not complete. Thus, dark fermentation results in, not only hydrogen, but also carbohydrates that require elimination because they are negative in the energy balance of the hydrogen-producing organism. This can be done by coupling one of the products of dark fermentation, acetate, to an additional container with a follow-up fermentation. This second fermentation is then a photoheterotrophic process using the organic acids produced in the first reaction and producing hydrogen and carbon dioxide (deVrije and Claasen, 2003)

6.3.7   Artificial Photosynthesis

Another interesting scientific and industrially potent idea is to produce hydrogen by exploiting solar energy directly. The total energy that reaches Sweden as solar energy is approximately 1000 times more than all the energy consumed in Sweden, which is 636 TWh (Swedish Energy Authority, 2006). In artificial photosyntesis (Hammarström and Styring, 2005), the energy is produced by splitting water into oxygen and protons, the protons being mediated by a hydrogenase, which catalyzes the production of hydrogen gas (Tamagnini *et al.*, 2002). The technique is based on the only molecule existing in nature that can split water to oxygen directly. This molecule is chlorophyll and is localised in the photosystems of the photosynthetic apparatus. As already mentioned, many microorganisms have, in addition, enzymes, called hydrogenases, that are capable of making hydrogen gas from protons and electrons, using an iron complex. The scientific task is to copy the central components of all these enzymes, in order to make artificial catalysts that will perform similar reactions.

6.4   Thermochemical Processes

Biomass can be converted to different energy products, such as heat and electricity, different transport fuels, such as methanol, DME and other chemicals by thermochemical processes. These thermochemical processes can be divided into several main groups:

- Gasification of biomass e.g. to different hydrocarbons such as methane.
- Pyrolysis i.e. heating of biomass in the absence of oxygen to produce different products.
- Directly liquefying biomass (liquefaction) by, for example, high temperature pyrolysis or high pressure liquefaction.
- Combustion in the presence of air to convert the chemical energy to power, e.g. heat and mechanical.

Biomass is mostly composed of cellulose, hemicellulose and lignin and some extractives. There are several factors that influence the choice of conversion process; the desired form of bioenergy, the type and quantity of biomass feedstock, and also the environmental and economical aspects. Biomass materials such as agricultural and forestry residues, etc., are often difficult to use directly in thermochemical conversion processes to different energy products due to their inhomogeneous and bulky nature and they should go through some kind of pretreatment or physical upgrading process as already mentioned in Section 6.2. Biomass often contains a large amount of water, which should be removed to some extent, in many cases, before thermochemical treatment occurs. Biomass is often crushed, dried and milled before any thermochemical process. The cumulative particle size distribution for biomass is often determined by different methods in order to, in turn, determine the quality of the biomass and to enable its classification for different purposes. Torrefaction, as already described in Section 6.2.5 is another pretreatment process which improves the thermochemical conversion of biomass, e. g. it allows more efficient gasification and less smoking during the gasification of wood.

6.4.1   Thermal Processing Equipment

There are several different types of thermal processing equipment for biomass, e.g. the fixed bed gasifier (sometimes also called the moving bed), the fluidised bed reactor (gasifier), etc. The simplest gasifier equipment is the fixed bed reactor, which consists of an upright cylindrical container that has an inlet and an outlet for the gases and where the feedstock is fed from above and the ashes are removed at the base of the container. There are several different constructions of fixed bed reactors, such as the 'updraft gasifier', the 'downdraft gasifier' and the 'cross-flow

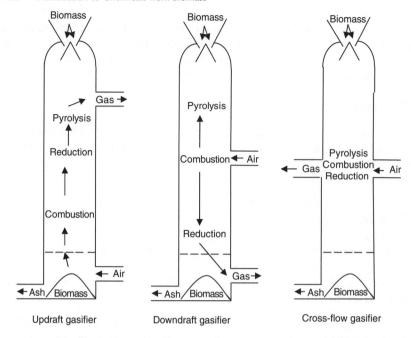

Updraft gasifier   Downdraft gasifier   Cross-flow gasifier

Figure 6.5   *Three different designs of fixed bed reactors for gasification*

(crossdraft) gasifier', see Figure 6.5. The updraft reactor permits the vapour of liquid compounds and tar to distill (flow) over, while for the downdraft gasifier the only way out for such materials is through the highest temperature zone, where they are destroyed. Therefore the amount of tar is minimised in this process and often any further purification of syngas is not necessary. In the cross-flow gasifier (reactor) the syngas is drawn off from the opposite side of the incoming gas, which prevents any contact of the reaction products with the fresh feedstock. Using this method the syngas is contaminated with a high amount of tar. There are also some other gasification reactors, such as the entrained bed, twin fluid bed, bubbling fluid bed and circulating fluid bed reactors that are not mentioned here, but are well described by Bridgwater, 2003. There are several pilot plants available for the gasification of biomass with different reactor techniques, e.g. the entrained flow technique in Germany (in Freiberg) and Sweden (in Piteå) and the circulating fluidised bed technique in Austria (Güssing).

Another type of gasifier is the fluidised bed reactor, based on a counter-current flow of gas and particulate material (fluidising medium) which is normally a kind of quartz sand bed with a controlled particle size (constant size distribution of particles). By this method a uniform temperature is maintained throughout the bed enhancing the overall yield of gasification at a lower temperature than is required in the fixed bed reactors. One drawback of this method is the high loss of sensible heat and a residue of both fuel and ash in the syngas. By using a cyclone it is possible to separate fly ash and particles from the syngas and there are several methods to overcome these problems. It is possible to summarise an efficient gasification of biomass in a fluidised bed reactor as follows:

- Biomass is fed into the fluidised bed
- Biomass is heated with hot sand (about 1000 °C) and syngas and char are formed
- The syngas and char are separated by use of a cyclone
- The syngas is further purified before its use as a raw material for fuel production
- The flue gas (the excess heat gas) may be used in steam production and power generation.

Ash related problems such as agglomeration occur, and exchange of the sand bed may be needed after some interval (Bridgwater, 2003). One way to overcome some problems with agglomeration is to use a mixture of biomass and peat in order to increase the melting point of the raw material in the furnace. It is also possible to use lime (contains CaO, Ca, etc.) instead of quartz in the fluidised bed reactor to improve the agglomeration temperature in the combustion atmosphere of some biomass (Natarajan *et al.*, 1998).

6.4.2   Gasification

In this process, carbon from biomass is converted to gases (CO, CO_2) by high temperature (above 800 °C). The produced CO_2 can react with hydrogen to directly produce methane, but also other different products, such as diesel, and other chemicals such as 1-alkenes in the presence of catalysts (Dry, 1999). This process has been used to produce Fischer–Tropsch diesel (FT diesel).

Fischer–Tropsch Diesel

Franz Fischer and Hans Tropsch (two German scientists) first studied the conversion of syngas (CO, CO_2 and H_2) into larger, useful organic compounds in 1923. Fischer–Tropsch (FT) diesel can be produced by gasification of coal or biomass. Coal or carbohydrates of different molecular length from biomass react with oxygen and steam to generate CO and H_2 via the Lurgi process. The mixture is called synthetic gas (or syngas or synthesis gas) and can be directly used in gas turbines for power generation and/or in the presence of different catalysts and different process conditions, converted to different fuels, such as methanol, DME, methane or FT diesel. The composition of the products is varied by different process parameters, such as temperature, pressure and residence time, and also by the kind of catalyst which is used in the process. One of the advantages of syngas production is the wide range of different feedstocks or raw materials that can give rise to syngas. When syngas is produced from a biomass precursor it is called biosyngas in order to distinguish it from syngas made from fossil raw materials. The raw syngas may be contaminated with sulfur compounds (e.g. H_2S), nitrogen compounds (e.g. HCN, NH_3), halides (e.g. HCl) and heavy organic compounds that are known collectively as 'tar'. There is also benzene, toluene and xylenes (BTX) in unpurified raw gas produced by gasification (Hamelinck *et al.*, 2004). These contaminants may inactivate the catalysts. Therefore it is often necessary to have a gas cleaning step for the raw syngas before conversion to different biofuels such as DME or methanol. Tar separation is a very complex and expensive process, but it is possible, for example, to remove some portions of the tar by an absorption step using a kind of oil in the purification step of syngas. There are several other methods for tar removal, e.g. thermal cracking, catalytic cracking and scrubbing methods, as mentioned by Hamelinck *et al.* 2004. Diesel is a mixture of aliphatic hydrocarbons (alkanes) with different carbon chains, i.e. it is usually a mixture of many hydrocarbons (C_{12}–C_{22}) but no aromatic hydrocarbons. A generalised example of the conversion of biomass carbon to different compounds, such as methane and other alkanes by Fischer-Tropsch method is shown below:

Lurgi process:

$$C + H_2O \text{ (Steam)} \xrightarrow{\text{Heat}} CO + H_2$$
$$\text{Syngas}$$

Fischer-Tropsch synthesis:

$$CO + H_2 \xrightarrow[\text{Heat, Pressure}]{H_2, \text{ Fe}} \text{Alkane} + H_2O$$

or

$$CO + H_2 \xrightarrow[\text{Heat, Pressure}]{2 H_2, \text{ Ni}} CH_4 + H_2O$$

The process requires high temperature (200–350 °C) and high pressure (10–40 bar). The carbon source for FT diesel has varied throughout the years; during World War II, coal was used by Germany to produce liquid fuel. South Africa was boycotted by most of the oil-producing countries and had to produce fuels and chemicals from coal by FT synthesis for several years. In South Africa, the production of several compounds, such as ethane and propene, for polyethylene (PE), polyvinyl chloride (PVC), polypropylene (PP) and acrylonitrile is by the FT-process (Dry, 1999). Nowadays, natural gas has been used to produce synthetic gas. Unfortunately, biomass has not yet been commercially applied as a feedstock for the production of synthetic gas. However, it is more interesting to use biomass as a CO_2 neutral source rather than as a fossil-fuel precursor. Biomass, on a laboratory scale, has been used to produce FT diesel, but there are several factors which make it difficult to produce FT diesel from biomass on an industrial scale e.g. gas cleaning (purifications) steps and scale-up of processes. FT diesel production can be combined with electricity production to increase energy efficiency (utilisation) and reduce production costs. FT diesel made from biomass is free of sulfur and nitrogen and contains no (or very few) aromatic compounds, which makes it more environmentally friendly than diesel produced from a fossil precursor.

6.4.3   Pyrolysis

By definition pyrolysis of a compound containing carbon is incomplete thermal degradation in the absence of oxygen, which results in liquid and gaseous products, and also char. The liquid product from biomass pyrolysis is known as bio-oil or pyrolysis oil (there are several other names available such as biofuel oil, wood liquid, wood oil, etc). Bio-oils are a mixture of different molecules (alcohols, aldehydes, ketones, esters and phenolic compounds) derived from the fragmentation of lignin, cellulose, hemicellulose and extractives. It is much easier to handle and to transport bio-oil than solid biofuel. However bio-oil can change during storage if the products in the condensate (bio-oil) have not reached thermodynamic equilibrium during pyrolysis. The moisture content of bio-oil is about 15–30 wt% of the original moisture of the feedstock is used in the pyrolysis process (Qi *et al.*, 2007).

Pyrolysis of biomass is divided into slow pyrolysis, which is well known to produce charcoal, for example, fast pyrolysis, which produces a high yield of liquid biofuels and other chemicals (Bridgwater, 2000) and flash pyrolysis. Slow pyrolysis (or carbonisation) requires low temperatures and very long residence time. In the carbonisation process the amount of char is maximised.

Fast pyrolysis of biomass occurs usually at 500–700 °C and high heating rates (e.g. 300 °C min^{-1}) over a short time, e.g. pyrolysis of pine wood samples at 550 °C results in the release of high amounts of aldehydes, ketines and methoxylated phenol. The most important products derived from pine by pyrolysis are turpentine and pine oil (sometimes called tall oil). Biomass should normally be dried to about 10% moisture content before fast pyrolysis (Yaman, 2004). Recently, biomass has been converted to bio-oil and then to hydrogen by catalytic steam reforming, however the yield is relatively low. Hydrogen can be produced by either gasification of biomass followed by reforming of the syngas or fast pyrolysis followed by reforming (rearrangement) of the carbohydrate fraction of the bio-oil. Hydrogen has also been produced by supercritical fluid (water) extraction of wood at different temperatures (Demirbas, 2007).

Flash pyrolysis is the process in which the heating rate is very high and the reaction time is of only a few seconds. Therefore, the particle size of biomass should be fairly small (105–250 μm) for this process (Goyal *et al.*, 2006). Often flash pyrolysis and fast pyrolysis are mentioned as one and the same process in the literature.

The gaseous products of fast pyrolysis require rapid cooling or quenching to minimise secondary reactions of the intermediate products (radical components). These radicals are very reactive and can undergo secondary reactions such as cracking and carbon deposition. The stabilisation of the radicals can be done by, for example, hydrogen addition (quenching). There is a range of chemicals that can be produced from bio-oil, including food flavouring agents, phenols, etc., by extraction and/or other reactions. Many examples of power generation from biomass liquids produced by fast pyrolysis processes have been reported (Chiaramonti *et al.*, 2007).

There are several parameters that have an effect on the yield and the composition of the volatile fraction of biomass during pyrolysis: the biomass species, the chemical and structural composition of the biomass, the temperature, the particle size, etc. (Demirbas, 2002).

Liquid production with high yield from biomass by fast pyrolysis is a promising technique for the replacement of fossil-fuel precursors for different chemicals and fuels with sustainable and renewable energy sources (Qi *et al.*, 2007).

6.4.4   Liquefaction

Liquefaction is a thermochemical conversion process in which liquid is obtained from biomass, sometimes called the biomass to liquid (BTL) process. In this process different liquid compounds are produced at low temperatures (300–350 °C) and high pressures (5–20 MPa) over a residence time of about 30 minutes and often using a catalyst in the presence of hydrogen, which is called catalytic liquefaction. But there is also a non-catalytic aqueous liquefaction of biomass, which is often called direct liquefaction. The yields and composition of bio-oil and char are different, depending on whether the catalytic or direct liquefaction process is used. Liquefaction and pyrolysis are often confused with each other, but there are several differences between them, e.g. pyrolysis usually occurs at higher temperatures and lower pressures than liquefaction and there is no need to dry the biomass before liquefaction, in contrast to pyrolysis, which often requires a predrying step. Interest in the liquefaction process is lower than the pyrolysis process due to the need for more expensive reactors and fuel-feeding systems in the liquefaction process. During liquefaction, a mixture of gas, liquid (bio-oil) and solid products is formed in varying proportions, depending on the different reaction conditions (Demirbas, 2000; 2001).

6.4.5   Combustion

Combustion of biomass (burning of biomass in air) is another alternative for converting chemical energy in biomass to heat and electricity. The combustion process is usually performed in a furnace, boiler, steam turbine, turbo-generator, etc. Combustion produces hot gases at temperatures around 800–1000 °C or more. The heat produced must be used immediately for heat and/or power generation, as storage is not a viable option. The combustion of biomass can be considered as several steps: devolatilisation to char and volatiles, and combustion of the volatiles and the char. There are several properties that are important for combustion of biomass, e.g. volatiles composition, tar and ash content, etc. (McKendry, 2002). Therefore the characterisation of different biomass feedstocks is necessary. The characterisation of biomass is divided into two categories: chemical and physical characteristics. The chemical characteristics are moisture content, ash content, ash-forming elements, concentration of other inorganic elements and ash-melting behaviour during combustion. These chemical properties depend on the type of biomass feedstock and the agricultural conditions. The biomass should sometimes be predried to have a moisture content of <50% before combustion. The physical characteristics of biomass feedstock after treatment by different processes include particle size, particle shape, bulk density, etc. There are several advantages to upgrading biomass before combustion, which have already been described in Section 6.2.

Combustion can be carried out on a small scale (for household heat production), on a medium scale (in hospitals, government offices, etc.) and on a large scale (in combustions plants for industrial purposes) (McKendry, 2002).

The effective combustion of biomass normally yields nothing more than CO_2 and H_2O. But if the biomass has been contaminated (treated) with chlorine, e.g. PVC-coated wood, together with carbon, hydrogen and oxygen, dioxins can be formed during combustion (mainly during cooling of the flue gas). Dioxins belong to a group of organic compounds (they contain aromatic rings, chlorine and oxygen), which may have some carcinogenic properties (Hedman, 2005).

6.5   Chemical Processes

Many of the existing energy products are being used as fuel for the transport sector and are made from either petrol-based material or from renewable resources such as biomass by some kind of chemical process. The most common biofuels that are being used for transport purposes are dimethyl ether (DME), methanol, ethanol, butanol and biodiesel.

6.5.1   Dimethy Ether (DME)

Dimethy ether, DME is the simplest ether that is a gas at normal temperature and pressure. DME is non-carcinogenic, but highly flammable. DME, with the chemical formula CH_3OCH_3, can be used as fuel in diesel engines, either as pure fuel or blended with diesel, because of its low degree of self-ignition. DME has a high cetane number, 55–60 (diesel has a cetane number of 40–55) and therefore has a shorter ignition time and better combustion than diesel. DME as a fuel is usually handled under pressure, in liquid form. Since DME can be produced from synthesis gas, it is possible to use many raw materials including biomass using the gasification method. However, most of the DME produced today is derived from fossil raw material. One common method is dehydration of methanol to DME. The use of DME as a fuel requires pressure vessels and a new distribution system, which would increase the distribution costs considerably.

DME has also been used as an additive to methanol to shorten its ignition time. DME has been used for other purposes, such as residential fuels for cooking and heating, power generation and also for hydrogen-rich fuel cells (Semelsberger *et al.*, 2006).

Other ethers, like methyl-*tert*-butyl ether (MTBE) and ethyl-*tert*-butyl-ether (ETBE) have been used as additives instead of aromatic compounds in fuels to increase the octane number of the fuels. However, MTBE and ETBE can be used as fuels by themselves in Otto motors as well.

6.5.2 Biodiesel

Biodiesel is a fuel made from natural (biological) renewable resources, which can be used directly in conventional diesel motors (engines). Biodiesel has several advantages compared to diesel produced from fossil precursors, for example it is degradable, non-toxic, contains no sulfur and releases less emissions during combustion (Zheng *et al.*, 2006). Biodiesel, sometimes called FAME (fatty acid ethyl ester), can be produced from a number of different raw materials, such as palm oil, soybean oil, rapeseed oil, sunflower oil and several others vegetable oils, by slight chemical modification – so-called transesterification. One negative aspect of biodiesel is that the quality changes with storage, due to oxidative and hydrolytic reactions. The other limitation is the availability of the raw material for biodiesel production. One of the most common kinds of biodiesel is made from rapeseed oil and is called rapeseed methyl ester (RME).

6.5.3 Rapeseed Methyl Ester (RME)

Rapeseed methyl ester (RME) is another alternative biofuel that can be used in diesel engines. RME has the advantages that it is renewable compared to diesel, non-toxic and less flammable compared with many other fuels, like ethanol. RME has the same cetane number, viscosity and density as diesel, contains no aromatic compounds and is biologically degradable with minor contamination in soil. RME can be produced from vegetable oils, but is mostly produced from rapeseed oil by pressing of the seeds or by extraction. Up to 3 tons of rapeseed can be produced from one hectare. The fatty acids in rapeseed oil are mostly oleic acid, linoleic acid and linolenic acid. The oil is pressed from the plant and after some purification allowed to react with methanol in the presence of potassium hydroxide as a catalyst, to produce a methyl ester, see Figure 6.6.

There are two by-products of this transesterification reaction: one is glycerol, which can be used in the cosmetics industry or can be used as a raw material for producing other industrial chemicals (e.g. 1,3-propanol or 1,2-propane-diol (propylene glycol)), and the other is a protein-rich cake which is left after pressing the oil from the seeds and can be used as stock feed (Saha and Woodward, 1997). These by-products offset the production costs of RME and make it a potential alternative to diesel. The energy output:input ratio for rapeseed biodiesel is between 1 and 6 (results from different countries in Europe). The European Union has a restriction on how many hectares each country is allowed to use for the cultivation of oil plant crops, e.g. in Sweden it is 120 000 ha, which can give around 80 000–100 000 m^3 rape oil or RME. Unfortunately it is not possible to cultivate rapeseed every year and several years (up to 4–6 years) of alternative cultivation of crops other than rapeseed are needed to avoid insect and fungal damage. It is also possible to use RME as

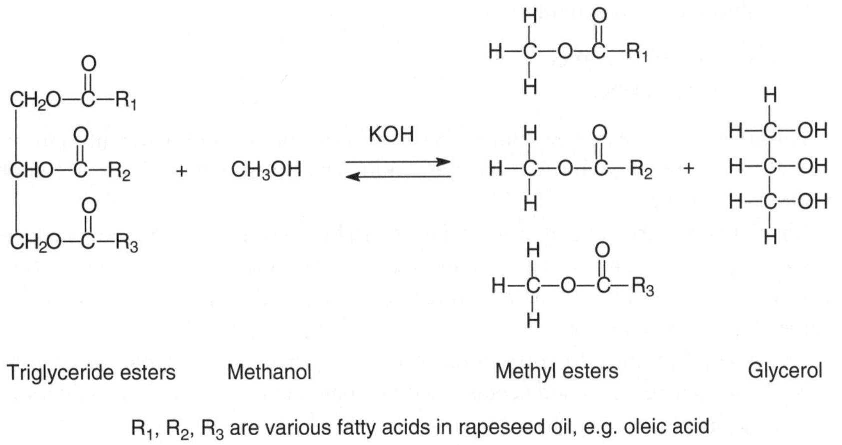

Triglyceride esters   Methanol   Methyl esters   Glycerol

R_1, R_2, R_3 are various fatty acids in rapeseed oil, e.g. oleic acid

Figure 6.6 *Transesterification of triglycerides to methyl ester and glycerol*

mixture with diesel. RME can be handled in the present diesel distribution system without any modification. Some rapeseed oil is also used in the food industry.

6.5.4 Primary Alcohols

Methanol, ethanol and butanol are different liquid biofuels that can be synthesised from biomass and can be used in both Otto and diesel engines. These alcohols can be prepared from sugarcane, sugar beet, wheat, barley, corn, raw sugar, switch grass, agricultural residues, wood and many other industrial wastes, and recently corn stover. The most important characteristic of alcohols, which makes them suitable as a fuel for Otto engines, is their high octane number. The octane number is a numeric presentation of the antiknock properties of a motor fuel. By definition the octane number is:

- $0 = n$-heptane
- $100 = iso$-octane (2,2,4-trimethyl pentane).

For other fuels the octane number is decided by comparison with a mixture of these two compounds.

With a high octane number it is possible to push a fuel–air mixture into the engine's cylinder (higher compression ratio gives higher efficiency and less fuel consumption) without any risk of uncontrolled self-ignition which may cause 'knocking' and serious damage to the engine as a consequence. Unfortunately, ethanol and methanol have low centane numbers (8 and 5 respectively) which means it is only possible to use them in diesel engines if some ignition improver (e.g. di-*tert*-butyl-peroxides) is added to them. These kinds of additives are usually costly. The cetane number is a numeric presentation of the fuel's ignition properties.

By definition the cetane number is:

- 15 = hepta methyl nonane
- 100 = n-hexadecane.

For other fuels the cetane number is decided by comparison with a mixture of these two compounds. Too low a cetane number causes slow ignition and poor engine performance.

Since it is possible to add at least 5% ethanol to gasoline without the need for any changes to the engine, this would dramatically reduce the net concentration of CO_2 in the atmosphere. Ethanol and methanol are highly flammable and have a flame that is difficult to see.

The ethanol produced can be used to make other related chemicals, such as ethyl acetate, acetic acid and acetaldehyde by chemical reactions, e.g. oxidation, esterification.

Methanol

Methanol is a colourless liquid without any particular smell at room temperature. It can be used as fuel in Otto and diesel engines and is highly toxic, corrosive and flammable, and since it can catch fire with an invisible flame it is not easily visually detected. However, waste or pollution of methanol in water or soil is degradable relatively quickly and does little harm to the environment. Methanol can be produced from fossil fuel and also from renewable resources such as biomass and other cellulose materials. It is usually produced from natural gas, but methanol can also be produced by gasification of biomass to a synthetic gas which, in the presence of catalysts, under high pressure and temperature (e.g. 220–275 °C, 50–100 bar and Cu/Zn/Al as catalysts) is converted to biomethanol containing some DME and water, which can then be purified by distillation. Methanol has been used as an oxygenate (high-octane oxygen-containing compound) as a blending agent in gasoline. It can be used neat and is called M100 or blended with gasoline, mostly as 85% methanol and 15% gasoline and called M85. Methanol is a bulk chemical (produced globally) that is used also in large amounts industrially. Methanol is an important chemical intermediate used to produce a number of other chemicals, including formaldehyde, dimethyl ether, methyl *tert*-butyl ether and acetic acid, for example. The energy content per litre of methanol is almost half that of petrol, but its octane number is much higher, making it an energy effective fuel for vehicles. The addition to gasoline of aromatic compounds such as benzene or toluene, which are carcinogenic, will increase the octane number of gasoline. However, it is possible to add methanol to gasoline instead to increase its octane number.

Ethanol

Ethanol is a colourless liquid that has been used in many ways as a chemical in the medical and food industries. It is soluble in water and can be used as a pure fuel or blended with gasoline or diesel. With regard to safety and environmental issues, ethanol is less toxic than gasoline, diesel or methanol. Ethanol contamination can be broken down by bacteria to carbon dioxide and water. It can be blended with gasoline up to 25% (common in Brazil since 2002 and called gasohol) and used in Otto engines. Diesel and ethanol are difficult to blend and the addition of an emulsifying agent is necessary for a stable, homogenous solution. Pure ethanol has been also used in heavy vehicles with diesel engines and as a mixture of up to 85% with gasoline (E85) in flexible fuel vehicles (FFV) such as the Ford Taurus. Emissions from the combustion of ethanol are much less than for fossil fuels, for instance emissions of particles, NO_x, CO, and other organic compounds. Ethanol can be produced by different processes and also from different raw materials. The raw material for its production can be cellulose, waste from paper mills, excess wine, in Europe, and also other biological materials. In contrast to methanol, ethanol can be produced biochemically by fermentation of carbohydrates (sugar) obtained from many different raw materials, such as sugarcane, sugar beet, corn, etc. There are three different feedstocks available for ethanol production, sugar feedstock, such as sugarcane, starch feedstock, such as cereal grains and potato, and cellulose feedstock, such as forest products and agricultural residues. Its production has increased in several countries, e.g. USA, China, Brazil, India and recently in some European countries.

Butanol

Biobutanol has many similarities to bioethanol and also some comparative advantages: it is easy to blend with gasoline in higher concentrations without any harm to engines, has a higher energy content (about 30% higher), is better able to tolerate water contamination and can be used together with ethanol and blended with gasoline. Biobutanol's low vapour pressure reduces the vapour pressure of the ethanol/gasoline blend and makes it easier to blend ethanol with gasoline. Biobutanol can be used in current existing vehicles without any modification to the engines, does not need any new distribution system and can be delivered by the existing fuel supply infrastructure. It can be produced by the fermentation of biomass and can use the same feedstock as bioethanol, e.g. sugarcane, sugar beet, corn, wheat. It has been shown previously that the sugars present in the hydrolysis step of biomass conversion could be fermented to butanol by *C. acetobutylicum*.

6.5.5 Ethanol from Sugar Feedstock

Ethanol from Sugarcane

Ethanol production from sugarcane (*Saccharum* sp.) is one of the easiest and most efficient processes since sugarcane contains about 15% sucrose. The glycosidic bond in the disaccharide can be broken down into two sugar units, which are free and readily available for fermentation. The sucrose is separated from the sugarcane by pressing the already chopped and shredded cane. The remaining solid from the pressing is fibrous and usually used as a fuel in the sugar mill. Several steps are involved in isolating sugar as a pure solid, including several crystallisation steps. Sugarcane production requires a tropical climate and Brazil has the largest sugarcane cultivation and was the first and biggest producer of bioethanol in the world. Actually, the cheapest bioethanol produced is from sugarcane in Brazil and the second cheapest is made from corn in the USA.

Ethanol from Sugar Beet

Sugar beet (*Beta vulgaris* L.) is a plant whose roots contain large amounts of sucrose. Sucrose is a glucose and a fructose unit bonded by a glycosidic bond, in another words, a disaccharide. Most of the European countries and Russia, together with the US produce most of the sugar beet in the world, e.g. the 10 biggest producer countries in Europe produced 242 million metric tons of sugar beet in 2005.

6.5.6   Ethanol from Starchy Feedstock

Ethanol from Cereal Grains

Ethanol production from cereal grains such as barley, wheat and corn is a much easier process than from cellulose material. The process includes several steps, as listed below:

- Milling of grains
- Hydrolysis of starch to sugar units
- Fermentation by yeast
- Distillation
- Removal of water from ethanol.

After grinding the raw material, it is mixed with water and enzymes to break down the starch to sugar units. The free sugar can be used by yeast or bacteria and converted to ethanol and carbon dioxide. As the concentration of ethanol increases to about 15%, fermentation is reduced, since high alcohol concentration kills the yeast or bacteria. It is then necessary to separate the ethanol from the other material in the fermentation tanks by distillation. Distillation increases the ethanol concentration up to about 95%. In order to remove the rest of the water from the ethanol solution it must be dried by different drying agents to a concentration of 99.9% ethanol, or absolute ethanol. The distillation and purification steps require a lot of energy and absolute ethanol will again absorb water from the air over time. There has been an ethanol company producing 50 000 m^3 ethanol per year from grain in Sweden since 2000. It is possible to produce 1 L of absolute ethanol from about 3 kg of wheat.

Ethanol from Corn and Corn Stover

One of the most common feedstocks for ethanol production is corn, which has been widely used in the USA. The starch in corn is converted to glucose after grinding in a dry mill, reacting it with dilute acid and then reacting it with amylases, e.g. α-amylase and glucoamylase. The free glucose is then available for fermentation to ethanol.

Some parts of the corn plant have been useless until now. Today, it is possible to use those parts of the plant, the stover, which includes stalk and leaves, as raw material for ethanol production. The amount of corn stover is huge, since for every kg of corn produced, almost the same amount of corn stover is left in the field to prevent soil erosion. Ethanol production from corn stover is more effective when enzymatic hydrolysis and fermentation are performed simultaneously, often called SSF, instead of having separate hydrolysis and fermentation stages (SHF). Corn stover mostly contains two carbohydrates: glucose and xylose. The amount of lignin varies between 17 and 26% of dry weight. Delignification of pretreated corn stover improves enzymatic hydrolysis, but is a costly step and may cause some loss of sugar units (Öhgren, 2006).

6.5.7 Ethanol from Cellulose Feedstock

Ethanol production from cellulose feedstock (often called next generation feedstock) consists of cellulose-rich biomass from many different precursors, e.g. wood and fast-growing plants like switch grass, reed canary grass, or crop residues from food production, such as corn stover. Ethanol production from lignocellulosic material includes two processes: (i) hydrolysis of cellulose and hemicellulose to different sugars; (ii) fermentation of the released sugars by yeast or bacteria. Ethanol can be produced from lignocellulosic material by chemical and microbiological hydrolysis processes. The chemical process is divided into two types, one using high acid concentration in the hydrolysis step, called the concentrated hydrochloride acid process (CHAP) and the other, developed in cooperation between Canada, the USA and Sweden (CASH) in which dilute acid is used in the hydrolysis step. In the microbiological process, cellulose is broken down to sugar units by cellulase enzymes. One big challenge in the ethanol production process is to avoid degradation of sugar to other organic compounds, such as furfural or 2-methy-hydroxyfurfural in order to keep the ethanol yield high (Wyman, 1996). These compounds act as inhibitors in the fermentation step.

A unique pilot plant for ethanol production from lignocellulose feedstock was inaugurated in Ö-vik, Sweden in May 2004. The aim of the pilot plant was to develop efficient continuous technologies for the various process steps in ethanol production from forest raw material and other lignocellulosic feedstock. Different raw materials require different conditions during the production process and the process also needs to be optimised for every raw material. Further it was important to demonstrate that large-scale lignocellulose ethanol production was possible and also to show that the production cost for cellulose-based ethanol could be decreased. The plant is complete with equipment to make it possible to do two-step dilute acid and/or enzymatic hydrolysis. In 2005, ethanol from wood chips (softwood, i.e. spruce) was produced in the plant. In the pilot plant it is possible to use 2 tons of biomass per day to produce about 400 L of ethanol. The lignin residue is used for other purposes rather than just heat production and power generation. Some research is going on concerning the use of the lignin as a precursor for producing other, more valuable, chemicals. It is possible to pelletise the lignin together with other biomass, e.g. in the Biofuel Technology Centre in Umeå, Sweden, lignin has been mixed with other feedstock to produce pellets for a more efficient transport of the lignin and the lignin has been used as solid biofuel. The cost of ethanol production will decrease dramatically when the lignin is used as a valuable product. The research results and experience gained from the pilot plant will be applied in full-scale industrial ethanol production from lignocellulosic feedstock in the near future. A schematic figure of ethanol production from a cellulose feedstock is shown in Figure 6.7.

CHAP Process

The CHAP process is based on the hydrolysis of lignocellulosic material by concentrated hydrochloric acid at low temperature and subsequent sugar fermentation.

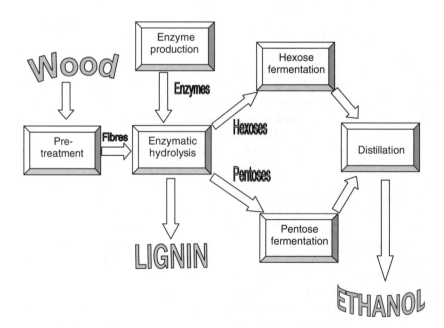

Figure 6.7   *Scheme of ethanol production from cellulose feedstock*

This process was developed for cellulose-rich raw materials since high concentration of the acid may cause degradation of the pentoses in hemicellulose to furfural derivatives. The ethanol yield is usually about 35%. However, corrosion problems and the need for higher capital investment and dangers associated with the recovery of the concentrated acid make this method less attractive. Furthermore, during combustion of lignin that is contaminated with hydrochloric acid there is some risk of dioxin emissions.

CASH Process

The CASH process was developed by cooperation between Canada, the USA and Sweden. In this method, hydrolysis occurs in two steps with dilute sulfuric acid at a temperature around 200 °C (pressure 8–25 bar) and the fermentation of sugars by yeast to ethanol. It has been shown that by using SO_2 and dilute sulfuric acid in two steps, this increases the sugar and ethanol yield, since the amount of inhibitors such as furfural is decreased. The process was developed for raw materials such as sawdust and other residues from trees. The ethanol yield is about 30% of the energy in the raw material and there are also by-products, with up to 40% of the energy content in solid form (lignin), which can be used as biofuel.

Ethanol from Switch Grass

Switch grass (*Panicum virgatum*) is a perennial C_4 species, which can be used in bioethanol production. This grass is grown in central USA as a fodder crop or for soil conservation. The lignocellulosic biomass has been used to produce ethanol with various yields. By pretreating the raw material to remove lignin and hemicelluloses, it is possible to significantly improve the hydrolysis of cellulose.

Ethanol from Reed Canary Grass

Reed canary grass (*Phalaris arundinacea* L.) is a perennial grass, about 2 m tall, with a sturdy, upright straw, broad leaves and a long panicle. The stem is partially surrounded by a sheath and is divided by nodes into shorter segments, called internodes. This grass grows naturally in Europe, Asia and North America, especially in wet and humus-rich soil. Reed canary grass consists mainly of cellulose, hemicellulose and lignin, but also proteins, lipids and a relatively high content of inorganic material. The main sugars after hydrolysis of reed canary grass are glucose, xylose and also arabinose; the amount of hexose in the stem varies between 38 and 45% of the dry weight of the material and pentose about 22–25%. The lignin content varies between 18 and 21% of the dry weight. The stem part of the grass contains more sugar than the rest of the plant and, today, there is a method available to separate stem and leaves fast and rationally. Therefore, the grass has very good potential as an alternative and a complement to short rotation woody crops and also to softwood feedstocks such as spruce, for ethanol production in the future.

Ethanol from Alfalfa

Alfalfa (*Medicago sativa L.*) is a feedstock for the production of fuel, feed and other industrial materials. Alfalfa consists mainly of celluloses, hemicelluloses, lignin, pectin and proteins. Therefore, it is a potential feedstock for ethanol production and also other chemicals.

Previous work has shown that it is possible to produce ethanol from alfalfa either by separate hydrolysis and fermentation (SHF) or simultaneous saccharification and fermentation (SSF). The yield of fermentable sugars from hydrolysis or saccharification is an important response variable in assessing the value of the feedstock.

6.6   Power Generation from Biomass

Electricity can be generated at the power station by several different methods, e.g. nuclear power or hydropower, and is distributed by a national or international network (grid) to the consumer. However during fuel production from biomass, huge amounts of steam and heat are often released as by-products, which can be used to produce electricity from renewable resources. In fact, many times the fuels produced have been used directly as an energy resource for power generation, e.g. to some extent, locally produced biogas has been used for heat and power production instead of using biogas directly in light or heavy vehicles. Therefore electricity from renewable resources should be considered as a green energy product.

Electricity production from biomass has several advantages:

- Low cost of biomass compared to fossil fuels.
- Security of supply and reduction of transport distances since biomass is often locally produced and some part of biomass can be used in electricity production.
- No net CO_2 emissions.
- Improvements of the economics of fuel production from biomass, since the electricity generated as a by-product can be used or sold and so reduce production costs dramatically.

Electricity production from biomass with fast growing conditions is usually integrated with other processes, e.g. pellet and ethanol production and also heat generation in a combined energy facility is an optimised and cost-effective method. Two interesting biomasses are corn and corn stover, since corn has been produced for food and also ethanol production in many countries and corn stover has been used in ethanol production and also recently pelletised with very good results in generating heat and power. Combining heat and power production (CHP) from biomass is a very common concept in many industries (Faaij, 2006).

6.6.1   Fuel Cells

A fuel cell is an electrochemical process for the production of electrical energy and heat. The concept is that chemical energy is converted into electricity and also heat without any combustion step. The process is clean (no emissions) and quiet and also effective compared to conventional combustion engines. The excess heat from fuel cells can be used to heat water. A fuel cell usually consists of an electrolyte surrounded (encased) by two electrodes: a negative electrode (anode) and a positive electrode (cathode).

There are several types of fuel cell depending on the working temperature and type of electrolyte:

- Alkaline fuel cell (AFC)
- Direct methanol fuel cell (DMFC)
- Polymer electrolyte fuel cell (PEFC)
- Phosphoric acid fuel cell (PAFC)
- Solid oxide fuel cell (SOFC) operating at high temperatures (500–1000 °C)
- Molten carbonate fuel cell (MCFC)
- Proton exchange membrane fuel cell (PEMFC).

In addition there is a biologically based fuel cell, the so called biofuel cell. (Davis and Hingson, 2007)

In the direct methanol fuel cell (DMFC), an aqueous solution of methanol is oxidised at the anode and reduced at the cathode.

$$CH_3OH + H_2O \longrightarrow CO_2 + 6\,H^+ + 6\,e^- (\text{amode})$$

$$1.5\,O_2 + 6\,H^+ + 6\,e^- \longrightarrow 3\,H_2O(\text{cathode})$$

The methanol concentration plays a significant role in keeping a stable power output. In a fuel cell, oxygen and hydrogen are released by oxidation of methanol producing electricity. The concept is shown in Figure 6.8.

The electricity generated in a fuel cell can be used to power some vehicles instead of direct combustion of other fuels. El-vehicles are quiet, more effective and produce no emissions compared to fossil fuel-using vehicles. For sustainable

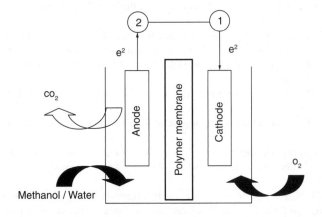

Figure 6.8 *Principle of operation of DMFC to produce electricity*

energy production, electricity used in cars should be produced from renewable resources. There are three types of vehicles using electricity as the energy source:

- Rechargeable battery vehicles, limited electricity,
- El-hybrid vehicles, one el-engine and one combustion engine,
- Fuel cells vehicles, one el-engine.

To conclude, fuel cells can be used in the next generation of vehicles as energy sources.

References

Adams, M.W.W. (1990). The Structure and Mechanisms of Ironhydrogenases. *Biochim. Biophys. Acta*, **1020**, 115–145.

Arshadi, M. and Gref, R. (2005). Emission of Volatile Organic Compounds From Softwood Pellets During Storage. *Forest Products J.*; **55**, 132–135.

Bridgwater, A.V. and Peacocke, G.V.C. (2000). Fast Pyrolysis Processes For Biomass. *Renew. Sustain. Energy Rev.*, **4**, 1–73.

Bridgwater, A.V. (2003). Renewable Fuels and Chemicals by Thermal Processing of Biomass. *Chem. Eng. J.*, **91**, 87–102.

Chiaramonti, D., Oasmaa, A. and Solantausta, Y. (2007). Power Generation Using Fast Pyrolysis Liquids From Biomass. *Renew. Sustain. Energy Rev.*, **11**, 1056–1086.

Chisti, Y. (2007). Biodiesel From Microalgae. *Biotechnol. Adv.*, **25**, 294–306.

Davis F. and Hingson S.P.J. (2007) Biofuel Cell, Recent Advances and Applications. *Biosens. Bioelectr.*, **22**, 1224–1235.

Demirbas, A. (2000). Mechanism of Liquefaction and Pyrolysis Reactions of Biomass. *Energy Conversion Manag.*, **41**, 633–646.

Demirbas, A. (2001). Biomass Resource Facilities and Biomass Conversion Processsing for Fuel and Chemicals. *Energy Conversion Manag.*, **42**, 1357–1378.

Demirbas, A. (2002). Gaseous Products From Biomass by Pyrolysis and Gasification: Effects of Catalyst on Hydrogen Yield. *Energy Conversion Manag.*, **43**, 897–909.

Demirbas, A. (2007). Progress and Recent Trends in Biofuels. *Progr. Energy Combust. Sci.*, **33**, 1–18.

DeVrije, T. and Claasen, P.A.M. (2003). Dark Hydrogen Fermentations, in *Biomethane and Biohydrogen. Status and Perspectives of Biological Methane and Hydrogen Production*, J. H. Reith, R. H. Wijffels and H. Barten (Eds). Dutch Biological Hydrogen Foundation, Smiet Offset, The Haag.

Dry, M.E. (1999). Fischer–Tropsch Reactions and the Environment. *App. Catal. A: General*, **189**, 185–190.

Faaij, A.P.C. (2006). Bio-Energy in Europe: Changing Technology Choices. *Energy Policy*, **34**, 322–342.

Goyal, H.B., Seal, D. and Saxena, R.C. (2006). Biofuels From Thermochemical Conversion of Renewable Resources: A Review. *Renew. Sustain. Energy Rev.*, **12**, 504–517.

Hamelinck, C.N., Faaij, A.P.C., Uil, H.D. and Boerrigter, H. (2004). Production of FT Transportation Fuels From Biomass; Technical Options, Process Analysis and Optimisation, and Development Potential. *Energy*, **29**, 1743–1771.

Hamelinck, C.N., Hooijdonk, G.V. and Faaij, A.P.C. (2005). Ethanol From Lignocellulosic Biomass: Techno-Economic Performance in Short-, Middle- and Long-Term. *Biomass Bioener.*, **28**, 384–410.

Hsu, T.-A. (1996) Pretreatment of Biomass, in *Handbook on Bioethanol: Production and Utilisation*, C.E. Wyman (Ed.), Taylor & Francis, London.

Hammarström, L. and Styring, S. (2005) Artificial Photosynthesis Towards Deeper Understanding of Photosystem II Function, in *Photosystem II: The Water/Plastoquinone Oxido-Reductase in Photosynthesis*, T. Wydrzynski and K. Satoh (Eds). Kluwer, Dordrecht.

Hedman, B., Burvall, J., Nillson C. and Marklund S. (2005). Emissions From Small-Scale Energy Production Using Co-Combustion of Biofuel and the Dry Fraction of Household Waste. *Waste Management*, **25**, 311–321.

Mani, S., Tabil L.G. and Sokhansanj S. (2006). Effect of Compressive Force, Particle Size and Moisture Content on Mechanical Properties of Biomass Pellets From Grasses. *Biomass Bioener.*, **30**, 648–654.

McKendry, P. (2002). Energy Production From Biomass (Part 2): Conversion Technologies. *Bioresource Technol.*, **83**, 47–54.

Melis A., Zhang L., Forestier M., Ghirardi, M.L. and Seibert, M. (2000). Sustained Photobiological Hydrogen Gas Production Upon Reversible Inactivation of Oxygen Evolution in the Gree Algae *Chlamydomonas Reinhardtii*. *Plant Physiol.*, **122**, 127–136.

Mohapatra, A. Leul, M., Mattsson, U. and Sellstedt. A. (2004). A Hydrogen-Evolving Enzyme is Present in *Frankia R43*. *FEMS Microbiol. Lett.*, **236**, 235–240.

Natarajan, E., Ohman, M., Gabra, M., Nordin, A., Liliedahl, T. and Rao, A.N. (1998). Experimental Determination of Bed Agglomeration Tendencies of Some Common Agricultural Residues in Fluidized Bed Combustion and Gasification. *Biomass Bioener.*, **15**, 163–169.

Öhgren, K., Rudolf A., Galbe M. and Zacchi G. (2006). Fuel Ethanol From Steam-Pretreated Corn Stover Using SSF at Higher Dry Matter Content. *Biomass Bioener.*, **30**, 863–869.

Olsson, M. (2006). Wheat Straw and Peat for Fuel Pellets – Organic Compounds From Combustion. *Biomass Bioener.*, **30**, 555–564.

Paulrud, S., Mattsson, J. E. and Nilsson, C. (2002). Particle and Handling Characteristics of Wood Fuel Powder: Effect of Different Mills. *Fuel Process. Technol.*, **76**, 23–39.

Prins, M.J., Ptasinski, K.J. and Janssen, F.J.J.G. (2006). Torrefaction of Wood, Part 1. Weight Loss Kinetics. *J. Anal. Appl. Pyrolysis*, **77**, 28–34.

Qi, Z., Jie, C., Tiejun, W., and Ying, X. (2007). Review of Biomass Pyrolysis Oil Properties and Upgrading Research. *Energy Conversion Manag.*, **48**, 87–92.

Robson, R. (2001). Biodiversity of Hydrogenases, in *Hydrogen as a Fuel: Learning from Nature*, Cammack, R., Frey, M. and Robson, R. (Eds.). Taylor & Francis, London and New York.

Saha, B.C. and Woodward, J. (1997). *Fuel and Chemicals From Biomass*, American Chemical Society, Washington, DC, 172–208.

Sedlak, M. and Ho, N.W.Y. (2004). Production of Ethanol From Cellulosic Biomass Hydrolysates Using Genetically Engineered *Saccharomyces* Yeast Capable of Cofermenting Glucose and Xylose. *Appl. Biochem. Biotechnol.*, **113–116**, 403–416.

Semelsberger, T.A., Borup, R.L. and Greene, H.L. (2006). Dimethyl Ether (DME) as an Alternative Fuel. *J. Power Sources*, **156**, 497–511.

Swedish Energy Authority (2006). *Energy in Sweden*.

Tamagnini, P., Axelsson, R., Lindberg, P., Oxelfelt, F., Wünschiers, R. and Lindblad, P. (2002). Hydrogenases and Hydrogen Metabolism in Cyanobacteria. *Microbiol. Mol. Bio. Rev.*, **66**, 1–20.

Wyman, C.E. (1996). *Handbook on Bioethanol: Production and Utilization,* Taylor & Francis, Washington DC, 1–424.

Yadvika S., Seekrishnan, T.R., Kohli, S. and Rana, V. (2004). Enhancement of Biogas Production From Solid Substrates Using Different Techniques – A Review. *Bioresour. Technol.,* **95**, 1–10.

Yaman, S. (2004). Pyrolysis of Biomass to Produce Fuels and Chemicals Feedstocks. *Energy Conversion Manag.,* **45**, 651–671.

Zheng, S., Kates, M., Dube, M.A. and McLean, D.D. (2006). Acid-Catalyzed Production of Biodiesel From Waste Frying Oil. *Biomass Bioener.,* **30**, 267–272.

Index

References to figures are given in *italic* type. References to tables are given in **bold** type.

Introduction to Chemicals from Biomass Edited by James Clark and Fabien Deswarte
© 2008 John Wiley & Sons, Ltd